CITIZEN SCIENCE IN BIOLOGY

CITIZEN SCIENCE IN BIOLOGY

The History and Development of Public Participation in Science

WILSON WALL AND DAVID MORGAN

PELAGIC PUBLISHING

First published in 2025 by
Pelagic Publishing
20–22 Wenlock Road
London N1 7GU, UK

www.pelagicpublishing.com

Citizen Science in Biology: The History and Development of Public Participation in Science

https://doi.org/10.53061/KLUQ2207

A CIP record for this book is available from the British Library

ISBN 978-1-78427-584-6 Hardback
ISBN 978-1-78427-586-0 ePub
ISBN 978-1-78427-587-7 PDF

EU Authorised Representative: Easy Access System Europe – Mustamäe tee 50, 10621 Tallinn, Estonia, gpsr.requests@easproject.com

Cover photo: A variety of Shark and Ray eggcases © Chloe Griffiths

Typeset in ITC Stone Sans Std by S4Carlisle Publishing Services, Chennai, India

Contents

Acknowledgements

It wasn't until we started writing this book that we realised just how immense the subject has become, and inevitably we have had to be selective with regard to the people and projects we have included. However, we would like to thank, and encourage, everyone who has ever taken part in citizen science, and we apologise to anyone who might feel that their contribution has been overlooked.

A huge thank you to the following citizen scientists who shared their personal stories with us: John Box, David Chelmick, Marc Grearson (on behalf of his father, John Grearson), Steve Trewhella, David Trudgill, David Wilkinson and, especially, Steven Falk. We are also grateful to the many friends who suggested citizen science projects, or individual scientists who they thought might add interest to the story we have tried to weave: Ian Brooking, Lindsay Davies, David Deisley, Dennis Doyle, Gillian Foulkes, Peter Foulkes, John Hawksford, Jackie Tasker and Robin Tasker.

In addition, we wish to thank everyone at Pelagic Publishing for recognising the value of this subject and for steering the book through to publication with such professionalism.

And, finally, of course, we are enormously grateful to our better halves, Alison and Val, for their forbearance and support during the whole process.

Introduction

A woman is watching sparrows from her living-room window. After a few weeks she begins to notice patterns in their activity, starts to take notes and is soon theorising about sparrow social behaviour. Elsewhere, a group of friends meet in a car park and commence wandering about, heads down, apparently aimlessly. They are botanists who are recording all of the wildflowers they can find in their chosen square kilometre of countryside. Their records will be combined with millions of other contributions and later used by specialists to study changes in species distribution.

All of these people are citizens with an interest in nature and the world around them. Only a very few of them will be professional scientists, but they are all citizen scientists doing citizen science.

There are many different views about what exactly constitutes citizen science. Since the term was first coined the number of definitions has proliferated, each one with its own unique emphasis. As a result, there are now well over 30 definitions. All of them originated after the appearance of the phrase in a paper titled 'Lab for the Environment', which appeared in an issue of the *MIT Technology Review* (Kerson, 1989). The article looked at three projects that involved environmental research and/or monitoring. However, it is not the projects themselves that are important

FIGURE 0.1 Citizen scientists in action: counting *Dactylorhiza* orchids to assess meadow restoration (see Chapter 4, page 81). (Graham Peet)

here, but rather the influence of this article and its creation of the concept of citizen science, as this has had considerable repercussions ever since. In fact this was not the first time the phrase 'citizen scientist' was used (although it is often claimed to be). However, it was the first time the term was used to describe the collection of mass data by citizens for scientific use, though obviously prior to this there were many examples of citizen scientists working in this way without being labelled as citizen scientists.

The *Oxford English Dictionary* defines a citizen scientist as follows: 'A member of the general public who engages in scientific work, often in collaboration with or under the direction of professional scientists and scientific institutions; an amateur scientist'. This is a useful catch-all definition because although the first part is often quoted, it is the second part, defining a citizen scientist as an 'amateur scientist', that is by far the most interesting. This is because all of the other definitions revolve around provision of data, not ideas, by the general public. Very often this involves a collaboration between scientist and layman in the hope of generating new knowledge and new interpretations of data. There is no doubt that definitions of citizen science can be highly variable, depending on the project and the people involved. Sometimes the citizen may be a passive provider of data, such as their blood pressure, while at the other extreme the citizen may be actively involved in collection and analysis of data. Although definitions help to give an overall idea of the nature of citizen science, we should not become bound by them. Projects are not just about big data, although that is a large part of modern methodologies. They are just as much about thinking, and individuals can do this very well indeed.

Citizen science has become something of a mantra among policymakers, who use the vast quantities of information collected to help to format ideas for the long-term public good, especially with regard to environmental and health policy. In the political arena the assumption that having more data is better is not always correct, and shows a lack of insight on the part of those who believe this is true. Simply put, questions can be raised about the validity of this argument, since – as every scientist knows – having more data of the same type does not necessarily lead to more reliable results, but may just lead to more complex calculations. Similarly, one might question whether many of the studies described as citizen science are really science at all. Although we can debate the exact meaning of the word 'science', we can generally agree that science helps us to understand the world in which we live, from atoms to planetary motions and all things in between. Simply collecting data can never of itself constitute science or scientific endeavour. Rather it is the reason for collecting the data and what is done with it subsequently that turns simple observation from natural history into science.

The other aspect of definitions of citizen science that raises a metaphorical eyebrow is when all that the individual supplies is additional computing power. For example, does searching data from radio telescopes for hints of extraterrestrial signals constitute citizen science or not? Is it simply an altruistic act by the individual who is helping in the same way as they would by donating to a charity? Sometimes we do need to look long and hard at projects that are described as citizen science and critically evaluate whether this is what they really are, or whether the label has been appended to a project for the benefit of grant-awarding bodies. Of course, branding a project as citizen science will also encourage citizens to sign up to it.

Sometimes it would seem that an entire industry has grown out of defining citizen science and creating ethical and scientific frameworks for originators of projects. The aim of most of these frameworks is to reduce the risk that the non-scientific participants will be exploited. This is all very useful for projects where large-scale data collection for analysis by computer and interpretation by a scientific panel is required, but what about the citizen scientists themselves?

A person whom we would call a citizen scientist and a dictionary would describe as an amateur scientist would traditionally have been referred to as a naturalist. Before the advent of formal training, and before there were paid professional scientists, virtually all scientific investigation was carried out by amateur scientists – that is, educated individuals with a strong curiosity about the world about them. For these natural historians, making progress in their particular field of interest was paradoxically both more straightforward and more difficult. The straightforward aspect can be demonstrated by the case of Isaac Newton. His question was a simple one to ask: 'What makes an object fall?' The difficult aspect was the need to express the results of experiments in a mathematical notation that was barely adequate for the task. The result was the introduction of a new unit, g, as an expression of acceleration. Nowadays it is the questions that are hard to formulate, and not necessarily the interpretation of the results. Before the twentieth century it was quite normal for scientists to work as individuals, as Newton, Michael Faraday and Gregor Mendel did. Since that time, collaborations have become the norm, and as these have increased it has become apparent that the biggest collaboration is with the general public. This opens up huge opportunities for the researcher.

Aldo Leopold, whose day job involved working for the Forest Service in Arizona and New Mexico, and who later bought a worked out, sand-blown farm in Wisconsin, was throughout his life a keen observer of nature. Importantly, it was the interpretation and understanding of what he witnessed that turned his observations into science, and in fact made him one of the founders of conservation science. Similarly, there are examples of contemporary citizens taking their observations to the level of science by asking questions of their results and doing relatively simple clarifying experiments that require little or only inexpensive equipment. These might be less well known and may not have full scientific rigour, but they can throw up some fascinating results and are incredibly enriching for the individual. Others are prepared to step even further into hard science and to invest in quite expensive equipment. Perhaps an extreme example is James Lovelock, a scientist by training and profession, who set up his own lab in his holiday home on the west coast of Ireland and spent his holidays measuring chemicals released by marine organisms. This was one aspect of his independent studies that fed into his Gaia hypothesis, according to which the Earth functions as a self-regulating system. These individual studies are fascinating but rare, as in the twentieth and twenty-first centuries collaborations have become the norm and, as mentioned earlier, the greatest collaborations are with the general public.

Just as there has been a proliferation of ideas as to what citizen science is, so there has also been a great deal of thought about what science itself is and how it moves forward. Although this can very quickly descend into philosophy and endless arguments over meaning, there are some points about science that are of real relevance and importance when considering citizen science.

Science tends to move forward on the basis of hypotheses that are tested. When a hypothesis has been tested to such an extent that there is no doubt about the outcome, the hypothesis becomes an accepted fact or theory. It is often imagined that a scientist will be working away at a project and have a 'eureka' moment when everything falls into place and there is another great leap forward in our understanding of the world. However, personal experience tells us that this is not generally how such things pan out – scientific understanding moves in a more haphazard way than this, but in the general direction of greater knowledge that allows us to modify our image of the world. The philosopher of science Thomas Kuhn suggested that science moves forward both in small steps, rather like puzzle solving, and in revolutions where a long-standing idea will stand as the model for research until so much contradictory information has accumulated that cannot be accommodated within the original theory that something has to give. Since the accumulating data are repeatable and reliable, if they do not fit the existing model the only option is to reformulate the original theory.

Small-scale or individual citizen science projects are usually of the puzzle-solving sort, whereas large-scale citizen science projects can be revolutionary, accumulating sufficient data to test hypotheses statistically and to throw up any discontinuities in data which may be of interest. The Search for Extraterrestrial Intelligence (SETI) project is like this. Assume that there are no extraterrestrials, but search for signals indicating that there are such beings. Analyse vast amounts of data and identify any anomalies that indicate there are intelligent beings out there. If a signal of this type was found it would induce a huge change in our attitudes to many different things. Such paradigm shifts can be seen in many other areas of science. For example, until the twenty-first century it was a truism that birds had feathers and reptiles did not, but accumulated evidence from the fossil record has now revealed that many non-flying theropod dinosaurs were feathered. As a result, our ideas and understanding of the Mesozoic Era have changed.

There has been an enormous proliferation of literature, this book included, on citizen science as a subject in its own right, the majority of such works being written from a social science perspective, but not this one. In twenty-first-century discussions of citizen science, projects that recruit large numbers of people to collect data are often described as 'democratising science'. This is a rather meaningless term which gives the unfortunate impression that the usual practices of science are elitist, or at the very least exclusive. The absurdity of many of the statements put forward when describing citizen science projects somehow misses the point that the participants, whether professional or amateur, are highly motivated and the projects aim to increase the sum total of knowledge while feeding the innate curiosity of everyone involved. Science by its very nature is open to everyone. Of course there are many areas that require a deep understanding based on a great deal of prior taught knowledge, which takes time, effort and determination to acquire. Perhaps it was easier in earlier times, when an untrained scientist or at least an unpaid one – a citizen

FIGURE 0.2 Jean Henri Fabre. A woodcut from a pen drawing by E. Heber Thompson.

scientist – could, with skilled observation, make notable contributions to scientific thought. There are numerous examples of this, many of which we will discuss later in the book. One such figure is Jean-Henri Fabre (1823–1915) who, though born into a very poor household, taught himself and became a schoolteacher, and in his spare time became a notable entomologist. His work on all manner of insect life, which was published in book form, has stood the test of time. Similarly, Gregor Mendel (1822–1884) not only taught himself plant breeding techniques, but also founded the whole scientific field of genetics and inheritance.

More recent times have seen a change of emphasis, and citizen science now tends to be focused on data collection rather than active participation in experiments. This is partly driven by the idea that the basic sciences, such as zoology, botany, chemistry and physics, are not well suited to direct citizen involvement because significant prior knowledge is required to understand the nuances of the subject. In contrast, data collection for conservation and environmental sciences is ideal for large-scale participation by the public.

It is often said that volunteers are involved in the generation of scientific knowledge, but in fact they are usually engaged in collecting data for scientific use. Of all the sciences that have utilised volunteers, either cogently or inadvertently, astronomy emerges as the first and most engaging. Early records of comets, equinoxes and the passing of the lunar months were kept and viewed as common knowledge among early civilisations. Fast forward to the twenty-first century, and a project called Galaxy Zoo used nearly 100,000 volunteers to distinguish between different morphologies of galaxies to aid astronomers' understanding of the formation and evolution of galaxies.

In the life sciences there are many biodiversity projects in which a picture can be built up not just of long-term changes, but also of the normal annual fluctuations that take place. These also lend themselves to long-term studies of phenology, which can give major insights into the impact of climate change. Studies of this type may span generations – for instance, the flowering time of Japanese cherry trees, records of which started in 801 CE and are still being kept today. Much later, but in a similar vein, Gilbert White in his *Garden Kalendar* kept records of events throughout the period from 1751 to 1773. Gilbert White worked on his own and, like many of the early citizen scientists, was independently funded, in his case by his work as a vicar. It was not unusual for such scientists to be individually sponsored, often by a Royal court, especially in mainland Europe. Many of these individuals were highly educated, though rarely in the fields to which they contributed. This suggests that we could define an individual citizen scientist as working outside their paid sphere of employment. According to such a definition, both of the authors of this book are citizen scientists, one collecting records for the Botanical Society of Britain & Ireland and the other being a tester of the AIDGAP keys for identifying invertebrates for the Field Studies Council.

Much of the observational data collected on plant and animal distributions is used either to test hypotheses or to help to describe levels of biodiversity in an area. Indeed, more than 80% of the biodiversity data in Europe is collected by all sorts of citizen science groups, from local observations of the presence or absence of species to international collaborations on subjects such as bird migration. For many of these projects the use of modern devices such as smartphones, bat detectors, trail cameras and small testing kits (to take simple measurements such as the pH of water) has increased by several orders of magnitude the monitoring of the biosphere. Apps such as iSpot, eBird and iNaturalist, among many others, make it very much easier to accurately identify and record where and when a species was seen. These are especially useful when studies are being made of groups that are intrinsically difficult to identify with certainty, such as flies and beetles.

In contrast to the projects in which individuals collect simple numerical data, there are far fewer experimental citizen science projects. The reasons for this generally revolve around the difficulty of supplying technical equipment to groups and individuals who will vary in their ability to use the kit effectively. However, when this issue can be overcome the field workers – our citizen scientists – have often gained insights into technical aspects of their project that are well outside their normal day-to-day activities. A very good example of this is the *Strålande Jord*, a project based in Sweden that encouraged school pupils to collect fungal fruiting bodies, weigh and identify them. The samples were then sent to Uppsala University, where radioactivity measurements were carried out. This project was designed to monitor the long-term effects of the Chernobyl nuclear disaster of 1986. Over 200 classes were involved in the survey, each of which was equipped with a 'research box'. The success of this was at least in part due to the organisers having learned from previous projects that a good, clear teacher's guide is essential for success. This crossover between the physical sciences, biology and radioactivity monitoring is quite unusual. Most of the successful large citizen science projects of this sort can be found in biology and conservation, such as biodiversity monitoring. Indeed, historically it has been citizen scientists who have made most species discoveries as well as recording and accumulating phenological data. In contrast, there have been very few projects in chemistry. Some citizen science projects use a certain amount of chemistry, but it functions as a tool, so the chemistry is effectively hidden from the participants. Examples of such projects include those that measure the pH of water or soil.

As has been pointed out, the competence and experience of the citizen scientist limit the amount of 'value-added' knowledge that there is for the participants. That is, the more the citizen scientist understands the project, the more they will get out of it. The results of projects seem to be much more reliable when the task required of the citizens can be quickly learnt and the participants feel responsible for the project. Citizen science projects in the life sciences generally consist either of large-scale collection of data on the same thing, such as repeated measurements for statistical analysis, or collection of data for construction of distribution maps that can be used to visualise changes in distribution of species or groups over time. Of course, in the second case some additional information is needed for the vital step of determining why a distribution has changed – for example due to climate, pesticides or shifts in land use.

Whether you are interested in large-scale collaborations or prefer to work on your own, the best thing to do is go out and observe, note down anything interesting or unusual and make yourself a working citizen scientist. We hope that this book will motivate you to participate and to help to make the world a better place through the sharing of experience and knowledge.

Chapter 1

The Early Years

If we think of citizen science as science being carried out by those working outside their area of employment, we can imagine that almost every early scientist fitted this description. However, there is more to it than that. A lone citizen scientist is essentially working in the dark, studying by observation, often with no background knowledge and hence with little or no foundation upon which to build. If we go as far back in history as we can, precisely these circumstances are found for everyone

FIGURE 1.1 Citizen scientists in action: Justin Paillot, nineteenth century French pharmacist, out botanising with his vasculum. See Chapter 2, page 94. (CC-BY-SA-4.0)

studying the natural world. There was no prior knowledge other than folklore and anecdotes, and the scientist would have to be a clear observer, rather than someone who casually assumed the truth of untested stories, such as the belief that maggots appear spontaneously from rotting meat. From the mid-twentieth century onwards the way that both formal and citizen science was carried out increasingly involved teams, rather than individuals, conducing the research and collecting the data. Nowadays it would be broadly true to say that the only time when a scientist is guaranteed to be working on a project singlehandedly is when they are conducting research for a PhD – the degree that recognises a person's research capabilities in a particular field of knowledge.

Science in the ancient world

Aristotle (384–322 BCE), often regarded as the first scientist in the Western world, was one of the early lone observers, who tried to make sense of what he saw and fit it all together to form a coherent body of knowledge. Aristotle was born in northern Greece and was for a while tutor to the young Alexander of Macedonia. He investigated many different phenomena and aspects of the world around him, and of course in his day there was no distinction between what we now regard as the different scientific disciplines – these were all regarded as part of the natural world. It is strange that Aristotle is best remembered as a teacher and philosopher rather than as a biologist, even though his primary contribution to the world was clearly scientific. His scientific reputation stems from two works, the *History of Animals* and *Dissections*. Unfortunately, the text of *Dissections* is not available to us, but from contemporary descriptions of it we can guess that, as suggested by the title, it consisted of illustrations and descriptions of the inner workings of the animals that Aristotle encountered. However, the *History of Animals* did survive in various forms that were transcribed by later scholars. The title is perhaps a little misleading, as the way that Aristotle uses the Greek word *Historia* refers more to research than to the passage of time or to studies of the past. There is similar confusion with regard to our term 'natural history', where again the word 'history' relates to research rather than to the past.

Aristotle's work stood the test of time because it was so comprehensive and rigorous by the standards of the day, and has not been surpassed for around 2,000 years. This is partly because few people were engaged in such investigations, but mainly because his work was so reliable. There are some mistakes and errors, but this is consistent with the fact that he was the first citizen scientist. As he was the first person to investigate many of the plants and animals that he described, there was no one from whom he could seek advice or a second opinion, and no existing body of knowledge to which he could refer. He was working without supervision and without any of the background information that we all take for granted. In view of these circumstances, what he achieved was astonishing.

Once background scientific information had begun to be laid down by observation it became easier for citizens to add to the burgeoning knowledge of the natural world. However, it would not be until long after Aristotle's time that people started to think of science as a profession. One of the reasons for this is that, if scientists are to be funded, the economy must be working well and structured so that scientific endeavour can paid for out of funds that may or may not be recovered from the results. This position was not achieved until centuries after the time of Aristotle, until which time private wealth or personal sponsorship was the only route for artists, philosophers and scientists to survive. It would not be until the nineteenth century that people would recognise what was meant when people described themselves as scientists by either profession or inclination. In the meantime it was the unpaid scientists, and specifically those scientists who were working and making a living in other professions, who were responsible for the great advances in our understanding of the world.

A good example of this is Ptolemy and his ideas about the universe. Born in Alexandria around 85 CE and living until around 165 CE, Ptolemy was an Egyptian Greek by birth, but historically it has been assumed that through one of his ancestors he was a Roman citizen, hence his name Claudius Ptolemy (Claudius is a Latin name and Ptolemy is a Greek one). Although he seems to have been primarily concerned with mathematical questions, Ptolemy is best remembered for his explanation of the geocentric nature of the solar system, with the Earth at its centre. This astronomical construction was embedded in some quite complicated mathematics which was described in his work titled *Almagest*, which has been constantly reworked and translated into many different languages over the centuries. His influence, along with widespread support for and confidence in his work, resulted in acceptance of this geocentric description of the solar system for about 1,400 years. The change from a geocentric theory to a heliocentric one, with the Sun rather than the Earth at the centre of the solar system, was a direct result of the work of Nicholas Copernicus (1473–1543). Although Copernicus had studied mathematical astronomy in Krakow, he made his living as Canon at Frombork Cathedral and as administrator and secretary to various senior figures in Poland. It was during his time as Canon that he slowly began to link his observations of the planets and stars to a model which was radically different to that put forward by Ptolemy.

The original description put forward by Copernicus was short and non-mathematical, and it was not until many years later, in 1543, that he published a complete explanation in *De Revolutionibus Orbium Coelestium*. By this time his original idea had become generally accepted, though it received quite a lot of criticism from Protestant reformers who preferred the literal explanation in the Bible despite the accumulating data which contradicted that interpretation. It was the gradual accumulation of data by Copernicus that resulted in such a massive change in perceptions of the solar system. This is an example of Thomas Kuhn's argument that accumulated data make a previously held theory untenable. Thomas Kuhn put forward his idea of scientific advancement in his book *Structure of Scientific Revolutions*, first published in 1962, in which he suggested that a theory holds until so much conflicting data has been obtained that the theory has to give way to a hypothesis that is a better explanation for all the observations, not just those chosen to support the original theory. Data that effectively scuppered the geocentric system included measurements of Mars that showed it apparently reversing direction across the sky. Such observations became more and more difficult to reconcile, and the physical and mathematical assumptions that were needed had to become ever more complicated compared with the simpler explanation put forward by Copernicus.

Although in many ways we can describe Copernicus as a citizen scientist, he had been educated in both mathematics and astronomy. So while he was earning a living in various different ways, it was the great freedom given to him by his occupation that allowed him to devote so much time to pursuing his ideas in cosmology. It is interesting that, like Copernicus, until well into the nineteenth century many if not most scientists were financially supported by their occupations working for the Church.

Citizen science in the eighteenth century

Long before the inclusion of the sciences in the educational curriculum in the nineteenth century, higher education at establishments such as Oxford University incorporated little in the way of contemporary knowledge. It seems that for centuries the pinnacle of educational achievement had been frozen by the hand of dogma as a litany of ancient Latin and Greek texts mixed with theology. This would naturally lead even the most scientifically inclined into a career in the church, bereft of the background knowledge craved by the scientifically curious. One recipient of such an ecclesiastical education was Gilbert White (1720–1793). White was born in his grandfather's vicarage at Selborne in Hampshire, and was associated with Selborne for a long time before he finally took up residence as vicar there in 1784. His entire working life, from his graduation at Oriel College, Oxford up until his death in 1793, was concerned with ecclesiastical matters, usually as a curate or dean. For this reason alone we can clearly see that his work in natural history could well be described as that of a citizen scientist.

The start of phenology

Although Gilbert White is generally best known for his *Natural History and Antiquities of Selborne*, published in 1788, his most significant contribution is found in *Garden Kalendar*, which consists of his diaries written between 1751 and 1773. In fact White considered his observations of natural history to be of greatest significance, and viewed his gardening calendar as merely a notebook in which he recorded what he did during the gardening seasons. So it has been White the citizen scientist's records of his observations of the natural world around him and the dates of events in his garden that have been of particular interest, especially in more recent years.

What White was doing in his own way was a form of phenology – the dating of natural events, especially among plants, for which he recorded, for instance, dates of first appearance of leaves and first flowering. Until recently these data were regarded as showing simple annual variations with the seasons, but now of course these studies are of particular interest as they indicate the effects of our changing climate. Gilbert White's *Garden Kalendar* gives details of crop production in his garden with reference to the weather as well. Extracting information from this to identify any trends is quite a tricky process, but can be done. Although White's records were never intended to be a scientific tool, they were generated by a scientific mind, which is precisely why more than two centuries later his data are so valuable. He managed to lay a foundation for citizen science by altering the way in which nature was observed. His theoretical contribution to biology was probably minimal, but he did answer many questions by recording accurate observations rather than making assumptions. This approach has been embraced in subsequent years, and has been pivotal to building up a picture of the natural world. White's observations extend further, through his view that the interactions between individual animals via a food web are of greater significance than their taxonomic relationships. This concept of interactions between organisms is almost a definition of ecology and – although that term would not have been known to White – it leads us to regard him as an early ecologist.

Robert Marsham (1708–1797) was both a contemporary and a friend of Gilbert White, and although he is a less well-known citizen scientist, it could be argued that he had a far greater influence than White. He was also elected a Fellow of the Royal Society in 1780. The Marsham family home is in Norfolk, and there in 1736 Robert Marsham started a family project that was to continue until 1958, and was called *Indications of Spring*. Marsham had a list of over 20 indicators for which he recorded the dates. These included: the first snowdrop flower; first swallow; first song of a migrant bird; first butterfly; first cuckoo call; dates when leaves first appeared on 13 different tree species; and the breeding activity of rooks, frogs and toads.

Large amounts of data accumulated between 1736 and their publication by the Royal Society in 1789. Interestingly, this was the year before Gilbert White's *A Natural History of Selborne* was published. The original data were presented as a table, which meant that it was very difficult to spot any trends, let alone long-term ones, in the data. On the other hand, at that time it was the seasonal changes from year to year that were of interest – climate change had not even been thought of. It would have been much more satisfactory for the data to have been displayed as charts, as would be done by Florence Nightingale later on (see page 19), but at the time graphical representation of data was still very much in its infancy. After the death of Robert Marsham in 1797 his family continued the project right up until 1958, when his descendent Mary Marsham died. From 1798 to 1810, Robert Marsham's son carried out the measurements, then his grandson, followed by the Reverend H.P. Marsham, Major H.S. Marsham, Reverend A.F. Marsham and finally Mary Marsham. Although the data are of interest for many reasons, one finding that is often singled out is the progressively earlier appearance of leaves on oak trees, advancing about eight days over 250 years. This progressively earlier production of leaves has been mirrored elsewhere in phenological data. As already mentioned, many other measurements were made by the Marsham family over the two centuries. All of the indicators that were recorded followed seasonal variations in temperature, but direct measurements of temperature were not recorded. There are some very good reasons for this, which may not seem obvious at first glance, but are related to the way in which temperature is measured. Whereas nowadays we can simply buy a thermometer and use it to take measurements, relying upon its accuracy, this was not always the case. Present-day methods of recording temperature were simply not available to early investigators of seasonal changes in temperature. Devices such as mercury-in-glass thermometers were available, but were very expensive and of questionable accuracy. Accurate measurement of temperature in degrees Celsius is carried out on an arbitrary scale as an interval variable, that is a scale without a true zero point which means we can only say that one temperature is higher or lower than another. Consequently, for the data to be reliable, any temperature measurement technique that the Marshams used would need have to have utilised the same instrument across the two centuries. Yet they were indirectly measuring temperature by consistent use of the same indicators, such as the flight of butterflies and the flowering of plants. In some respects direct temperature records are not the most relevant ones, as it is not just the temperature rise over the centuries that is significant, but also the effect that it has on biological systems, and this is what the Marshams were measuring.

An easier way to conduct a phenological study would involve simply recording the date of a specific event, such as the flowering time of a particular group of trees or a single species of tree. This is precisely what has been done for the cherry trees of Japan. At first glance it would appear to be a spectacular example of citizen science data collection. Records of the time of flowering of cherry trees in Kyoto, the old capital of Japan, have been collected since 801 CE. In contrast, in ninth-century Britain, Viking invasions were taking place and what was yet to become a unified state was in political and military turmoil.

For the majority of the time that cherry flowering times were being noted, these records were not collected in any formal way, so the data were not uniformly gathered with regard to time and place. What was fortuitous and makes the data so valuable was the fact that Kyoto was the ancient capital of Japan until 1868, and as a result the data were inadvertently collected by many high-ranking officials and merchants in the form of casual comments such as 'today I saw the first cherry blossoming' in their journals and diaries, and these have been preserved across the centuries. Once the data had been pooled and analysed it became apparent that at least two cherry tree species were involved. The documents available indicated that until the mid-nineteenth century the trees were *Prunus jamasakura*, but from then on a cultivated flowering cherry, *Prunus × yedoensis*, became the most widely planted cherry tree. These two species are in full flower at slightly different times, and it was the very short period of a couple of days when they are in full flower that was taken to be the significant date for the time of flowering. By carefully adjusting the flowering times to historical events so that the date of flowering was correct, the data became a statistically usable and an informative reservoir of information. Perhaps the most interesting result of this inadvertent citizen science project was the close association of the times when the cherry trees are in full flower with sunspot activity.

Projects of this sort demonstrate the challenges of using randomly collected historical data from citizen sources when there was no specific requirement for precision at the time when the observations were made, especially with regard to biological phenomena such as flowering. Therefore the relevant data become buried among many other kinds of information. A determined approach is needed to turn such information into a usable form, as two scientists, Aono and Kazui, did with the data on flowering times of the Kyoto cherry trees by relating flowering time to temperature. Such projects also demonstrate that although we might consider a routine observation to be of little intrinsic value, combination of that observation with others of the same sort can be of huge and unforeseen value later on, especially if the recording of regularly taken measurements is seen as both possible and of interest.

Citizen science in the nineteenth century

The nineteenth century was a fertile time for citizen science, as this was when the transition from amateur to professional science was occurring. It was a time when the educated would have acquired at least some background knowledge of science, not just natural history. Education was changing in response to the industrialisation of nations, and the new sciences were everywhere, influencing daily life. However, the amount of science taught in schools was, by modern standards, wholly inadequate, despite the desire to be seen to be familiar with the workings of such developments as the Industrial Revolution and mechanisation that ranged from ships and trains to manufacturing. This situation has parallels in the twenty-first century, when constant use of a mobile device gives the impression that the user is familiar with the workings of electronic technology.

In France, similarly, advances in science and a thirst for knowledge were the backdrop for a naturalist who stands out as a citizen scientist and extraordinary observer. Jean-Henri Fabre (1822–1915) was born and brought up in a poor household in Aveyron, France, but his family's poverty did not prevent him from getting a good education. Unusually, he was largely self-taught, and by the age of 19 he had gained a teaching certificate and taken up a post at Carpentras where he taught physics, chemistry and botany. In 1849 he took a position in Corsica before moving back to the mainland at Avignon. During this period he was working as a science teacher but spending a great deal of his free time as an observer of natural history, particularly insects. Fabre kept records of his meticulous observations and wrote books about them. His writing style was often criticised by established scientists of the day for its anthropomorphic descriptions of insects. Despite the somewhat unusual way in which his knowledge was delivered, Fabre was an original observer and a genuine citizen scientist. It is a testament to his work in entomology that he is best remembered for this work, rather than as a teacher. His books were often written in narrative form, in a similar manner to Gilbert White's *Natural History of Selborne*. The difference was that children were Fabre's target audience, and although the information was sound it was viewed as superficial because of the way in which he delivered it. Despite these contemporary criticisms, we can get a good idea of his fame within his lifetime from a simple photographic portrait by a French photographer called Nadar (whose given name was Gaspar-Félix Tournachon). Nadar photographed such well-known figures as Baudelaire, Sarah Bernhardt, Debussy, Dumas (père), Liszt, Zola and Fabre. Although Fabre was famous in France, he only came to prominence in the English-speaking world when his works were translated at the beginning of the twentieth century. Although he was a great observer of the natural world, he never took the step of making generalisations about his work that could be tested, so he never took the next step towards a better understanding of biological systems.

Gregor Mendel

In contrast, Gregor Mendel (1822–1884) was a citizen scientist who used observations made from his own experiments to create a brand new science, namely genetics. Mendel was born on the family farm in Silesia in what was part of the Austrian Empire and is now the Czech Republic. During his early years he worked as a bee keeper and gardener, but he seems to have been dogged by bad health both in childhood and later on. This was compounded by financial problems in paying for his education. His precarious financial position seems to have been part of the motivation for him to join a monastery, as this not only provided him with a stable career but also gave him access to a free education. Certainly his own testimony indicates that following this path eased his day-to-day concerns about survival. He became a monk at what was formerly known as Brünn, now Brno in the Czech Republic, where he trained as a priest. Over the following years, although he was teaching on a day-to-day basis, he twice failed the examination to qualify as an official teacher. Despite this, he travelled to the University of Vienna as a student sponsored directly by the Abbot of the monastery. When he returned from the university he became a teacher, primarily of physics, and by 1867 he had become Abbot of the monastery.

It was probably his studies in physics that influenced the way in which he approached the study of inheritance. The nineteenth century was a time of many original discoveries in physics (Alessandro Volta giving his name to the volt, James Joule giving his to the joule, Georg Ohm giving his to the ohm, and so on), as new and exciting rules and laws of physics were discovered, making humanity the master of the physical world. This had the effect of engendering a rather arrogant reductionist attitude towards the biological world – surely it was possible to formulate rules for biology in just the same way as they had done for physics and chemistry. It was against this backdrop of the new sciences being invoked to explain all aspects of the world in which we live that Mendel approached the old conundrum of how characteristics are inherited. Until the work of Mendel it had been widely assumed that characteristics were inherited from parents by a simple kind of blending. This idea was reinforced by the observation of family resemblances and racial stereotypes, but there were no data to demonstrate whether this was a true observation or just an assumed association. Basically nothing was known about inheritance, but there were a lot of unfounded assumptions.

The problem with the way that inheritance was investigated was that it was based on livestock breeding. This involved breeding towards a goal, from the perspective of 'what we want' rather than an observational starting point of 'what happens if?'. So if an animal breeder wanted a better milk yield from a cow or more wool on a sheep, they looked at the cows that produce the highest milk yield, or the woolliest sheep, and bred their stock from those animals. This meant that they were effectively chasing desirable characteristics while having no idea how these were transmitted. Surprisingly, perhaps, it was not until more than a century after the work of Mendel that it was realised that for many characteristics the very large number of genes involved in inheritance of a single feature – as well as environmental considerations such as diet – does actually result in an apparent blending of characters. The simplest example of this apparent blending can be found in human hair colour. If you line up a random collection of individuals, so-called brown hair is clearly on a spectrum of imperceptible change from blonde to black. These are mostly genetically controlled with environmental influences in the form of bleached and dyed hair as well as environmental factors, like age and stress, causing hair to naturally turn grey.

Thus whereas livestock breeders were trying to control the slippery systems of multi-gene inheritance without understanding what was controlling their desired traits, Mendel stood back and took a reductionist look at what was happening when characteristics were inherited down the generations. This very first step was hugely significant in itself, but what came next was even more important. He looked for characteristics in his chosen experimental species – in this case Pea *Pisum sativum* – which were either present or absent. It is not always easy to determine the nature of the characteristics so that one can be confident one has put the plants in the right category. Perhaps the best way of demonstrating this is by listing the seven characteristics that Mendel used for his research. These are in pairs, each member of the pair being controlled by a single gene (either dominant or recessive).

1. **Shape of the ripe seed.** This could be round or 'roundish' with only shallow depressions (if any), or irregular and deeply wrinkled. Round is dominant.
2. **Colour of the cotyledons.** Mendel recognised three different shades – pale yellow, bright yellow/orange or bright green. Pale yellow is dominant.
3. **Seed colour.** Mendel described the colour as white or grey, grey brown or leather brown. This was a particularly interesting characteristic to examine, as he associated white seeds with white flowers and brown seeds with coloured flowers. In this way he had inadvertently managed to demonstrate a genetic phenomenon called linkage, whereby two characteristics tend to be inherited together, although they are controlled by different genes.
4. **Shape of the seed pod.** This could be either inflated or contracted and wrinkled, with the inflated form being dominant.
5. **Colour of the unripe pods.** These could be either green or yellow, with the green form being dominant.
6. **Position of the flowers.** The flowers could be borne either at the top of the stem, or along its length, with the latter position being dominant.
7. **Length of the plant stem.** This could be either about 0.5 m or 2 m, with the longer stem being dominant – although crosses did produce stems of intermediate length.

When various crosses were made between different plants that also differed in their appearance, these seven characteristics of the pea plants gave rise to well-defined ratios of progeny which clearly demonstrated that at least some of inheritance could be due to a single gene. Mendel gave us the terms *dominant* and *recessive*, as well as the notation by which these are written in genetic analysis. More importantly, though, he used his numerical and statistical skills to generate ratios of offspring that gave us new ways of looking at breeding experiments. We do not know whether Mendel had any further insights into the implications of his work for other aspects of heredity, but he was the originator of modern genetics with its concept of discrete units of inheritance, each of which controls a single characteristic. His scientific work was originally published in the *Proceedings of the Natural History Society of Brünn*, but went largely unnoticed until it was translated and republished in the *Journal of the Royal Horticultural Society*. While he was carrying out this research that was to change the way we understand inheritance he continued to work as a monk with responsibilities as a teacher. We also know from the *Proceedings of the Natural History Society of Brünn* that Mendel presented the meteorological records for the area to the Brünn Natural History Society. He was truly a citizen scientist.

Florence Nightingale

Another individual who might seem just as unlikely as Gregor Mendel to be involved in the story of citizen science is Florence Nightingale. Her contribution to scientific knowledge was so far removed from her daily life that she is better known for her work than for her science. She is often overlooked as a citizen scientist, as are the statistics that she collected, because of the publication of a very influential book, *Eminent Victorians*, by Lytton Strachey in 1918. This contained biographical sketches of Cardinal Manning, Thomas Arnold, General Gordon and Florence Nightingale. It can be assumed that these were arbitrary choices, since at the time when the book was written such individuals as Darwin and Brunel had already generated a far greater legacy than Manning, Arnold and Gordon. The book's author had studied history and literature and, we can assume, was neither interested in Nightingale's ground-breaking statistical data nor equipped to understand them. Among the

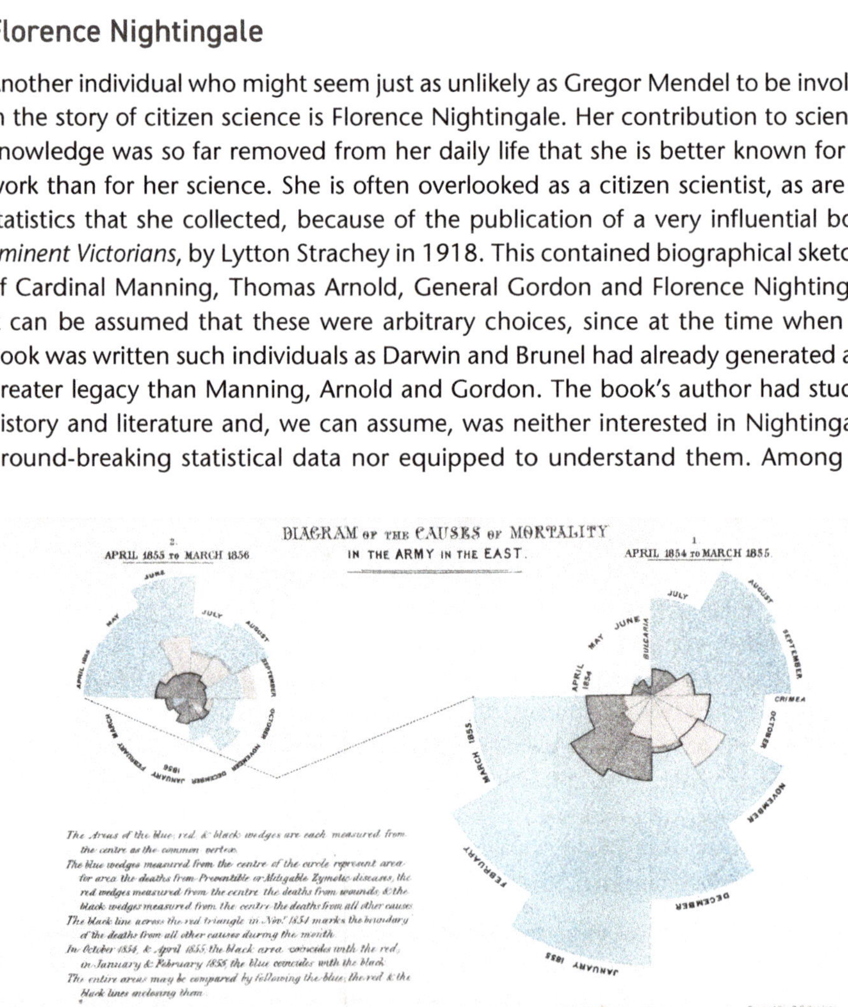

FIGURE 1.2 One of the polar pie charts that Florence Nightingale produced to display pictorially the mortality rates among soldiers in the Crimea.

innumerate literati at that time such a misplaced assumption of superiority based on their education was not uncommon.

As a result, Strachey's biographical account of Florence Nightingale focused on political machinations, both military and medical, and glossed over her meticulous collection of data on mortality rates and causes of death in the Crimea. It did not address the fact that when she was critical of the War Office she supported her arguments with data that she had collected and which she presented in chart form – a very novel technique at that time. Because of the influence of Strachey's book, the myth was promulgated that Nightingale's legacy was all about nursing, and about professionalising the subject while at the same time being extremely popular with the general public. However, in addition to that she was one of the first people to use polar charts to present data in graphical form. She collected her data in her personal notebooks during the Crimean War of the 1850s.

In 1829 a French lawyer and amateur statistician called André-Michel Guerry devised the polar chart as a new method of presenting data. This technique is well suited to recording annual events, as the angular sections remain the same but the radius varies in proportion to the numerical value each month. Polar charts are ideal for illustrating the month-by-month changes in death rates and infection rates that Florence Nightingale was trying to depict, because as the rate changes, so does the radius of the monthly arc. This resulted in a clear visual demonstration that at the hospital in Scutari (now Selimiye Barracks in Istanbul) there were ten times more soldiers dying from various diseases than were dying from their wounds. Until the nineteenth century data had always been presented in tabular form (as, for example, Robert Marsham had done when publishing his phenology data in 1789). This was as much due to a lack of graphical techniques as it was to a lack of coverage of statistics in education. Although mathematics had a long history in education, statistics was seen as a relatively minor subject, even though probabilities were used daily for all kinds of betting.

Despite lacking any formal scientific or mathematical training, Florence Nightingale recognised the need not only to collect data to support her argument for the necessity of more resources, but also – even more importantly – to present those data in a way that was not open to misinterpretation. This required her to learn a new method of presenting statistical data at a time when such techniques were still very much in their infancy. This was clearly recognised in her own lifetime, as it was only in 1834 that the Royal Statistical Society was founded, and Florence Nightingale was their first woman member.

Mary Anning

Slightly earlier than Florence Nightingale another amateur scientist was working alone, advancing our understanding of the pre-human ancient world. This was Mary Anning (1799–1847), and the nascent science with which she was involved was palaeontology. Mary was ten years old when she sold her first fossil, which she had found on the beach at Lyme Regis, Dorset. It is said that she sold it for half a crown (12.5 pence in decimal money, or adjusted to modern values, about £30). Her father was a cabinet maker who made extra cash by selling fossils from the blue lias and mudstone of the area to tourists. At a time when the middle classes were starting to go on holiday more and the desire to travel had increased, the French Revolution was making travel to France both more difficult and hazardous, so English coastal resorts became popular holiday destinations.

It was from her father that Mary learned the rudiments of fossil hunting, and how to wait for the winter storms and high tides to wash the cliffs away, causing landslides that exposed the embedded fossils. It was because of these landslips that fossil hunting on the cliffs was correctly regarded as very dangerous (indeed a landslip caused the death of Mary's dog, Tray). She was known as the most significant professional fossil collector of the nineteenth century, and many very important finds were attributed to her. Although she was the primary collector, the presentation of her finds to a wider public was usually carried out by her personal friends, Henry De La Beche (1796–1855) and William Conybeare (1787–1857).

One of the most significant finds that Mary excavated was the first ichthyosaur. This fossil had originally been found by her brother, Joseph Anning, but he did not have time to complete the excavation. After the fossil had been removed from the cliff and cleaned it was sold to a collector for £23, and then in 1819 it was sold on to the British Museum for £24. Only a few years later Mary – now a professional

FIGURE 1.3 Statue of Mary Anning with her dog, Tray, on the sea front at Lyme Regis. This is an elegantly stylised statue, as contemporary images of Mary portray her in floor-length dresses.

but untutored citizen scientist – excavated the complete skeleton of a plesiosaur, which was sold in 1823 for £110. These were highlights of her career, but there were many other finds which together led to a considerable shift in understanding of the surrounding fossil beds. Prior to these discoveries the status of fossil material was hotly debated, and questions were asked as to whether fossils were the remains of animals, and if so whether those animals were even extinct. After the extensive work of Anning and the descriptions put forward by Beche and others it was realised that what was being looked at was not a random collection of bones, but an ancient ecosystem, which repaid study for its own sake.

It was De la Beche who took this idea of an ancient ecosystem to its logical conclusion in a painting called *Duria Antiquor* (*A More Ancient Dorset*), which depicted various fossil species as they were envisaged to be in life, swimming and feeding. It was painted in 1830 and now resides in the National Museum of Wales, but before it was moved there De La Beche commissioned a well-known printmaker by the name of George Scharf to produce lithographs based upon the original which were sold privately for £2 10s (£2.50 in decimal money) to raise funds for Mary Anning, whose health was failing. The sum charged for the print was more or less the same as a contemporary skilled craftsman would earn in about 12 days. At about the same time, an annuity was raised for Mary by the British Association for the Advancement of Science and the Geological Society. After her death she was buried in St Michael's Church in Lyme Regis, sharing a headstone with her brother Joseph. In recognition of her work the Geological Society commissioned a stained-glass window for the church, and a local road was named Anning Road. Although she was basically untutored, Mary was highly skilled in spotting and interpreting the fossils that she

FIGURE 1.4 The station for cliff erosion photographs to be taken at a site in Lyme Regis.

saw. For instance, she was able to tell that fossilised material inside the stomach cavity of some of the larger species was a direct indication of a carnivorous diet.

The continued interest in fossils from the south coast of England stems in part from Mary Anning's work, but is also due to the fact that erosion of the cliffs brings a constant supply of new material to the surface. Monitoring of the erosion of the cliffs, which may be quite a slow process, has been helped by inviting passers-by to use their camera phones to record it. As can be seen in Figure 1.4, a group has set up a station where individuals can align their phones to take a photograph of a particular section of cliff in Lyme Regis. Each image can then be sent to the address shown on the plaque, allowing a composite image of change due to erosion to be created.

The rise of Royal Societies

To fully appreciate these individuals and their achievements we need to consider them in the context of the times in which they lived. It was during the nineteenth century that the Industrial Revolution reached its zenith, giving the technologically inclined and educated population a much greater sense of confidence and ownership of the world about them. This has often been interpreted mainly in terms of the effect on colonial expansion, but the true sense of ownership was in measurements – in the standards that gave scientific endeavours their first sense of being international. Units were becoming standardised, as were objects involved in measurements. For example, it was in 1839 that the Royal Microscopical Society, which started as the Microscopical Society of London, standardised the size of a microscope slide to dimensions that are still used today. Similarly with the threads holding the microscope objective lens to the microscope body, a small standardisation that has echoed down the years and made life very much easier for all of those professions that routinely use microscopes by making it easier to exchange lenses and slides between microscopes from different manufacturers.

Scientific societies were blossoming during this period and were often far more egalitarian than one might imagine. The Royal Entomological Society, for instance, was founded in 1833 as the Entomological Society of London and has always accepted female members, as has the Royal Statistical Society. The Royal Society of Chemistry was established in 1841 as the Chemical Society of London, gaining a Royal Charter after seven years, and the Physical Society of London, which was founded in 1874, was renamed the Institute of Physics in 1920. Although this society gained a Royal Charter in 1970, it has continued to be known as the Institute of Physics. The rise of the learned society can be seen as a response to the increasing literacy and numeracy of the population, which created a growing market for printed works. One area that was of considerable interest to these new readers concerned descriptions of the world at large – not just tales about the exploration of foreign lands, but also an explanation of the world around them in a manner that they could easily understand. With the rise of the logical, comprehensible science writing there developed the attitude that if a question could be asked that had not yet been answered, this was something that the individual could address. The developing logical aspect of scientific enquiry was due to the rational approach to the natural world described by Francis Bacon (1561–1626), building on the work of even earlier scholars. It involved new ways of thinking and new approaches to measurement – not just collecting measurements, but thinking about how measurements are made, and the best way of recording what had not been recorded before. As we previously mentioned, thermometers, for example, were rudimentary and not always directly comparable with each other. Until this time, many fields of study were regarded as complete, ancient texts being regarded as true, regardless of their content. This made for a very restricted education system which did not encourage the individual to question information given in classical texts.

We can see from this that early citizen science was carried out by individuals who were genuinely working in one area and investigating in another. These were commonly autodidacts, training themselves to make observations and measurements that had often not been done before, specifically to answer a question that they themselves had formulated. While doing this they may well have devised new methods of analysing data, or new methods of measurement, or in extreme cases they may have given rise to a whole new subject, as in the case of Gregor Mendel pioneering modern genetics on the basis of simple investigations of the growth habit of peas. It has been suggested that these lone individuals of the nineteenth century were not citizen scientists but amateur naturalists with private means and a dilettante attitude to science. Associated with that suggestion is the idea that real citizen science started with the large-scale involvement of the general public in the late twentieth century. This is not so – collection of data by the public started well before that, although whether you can regard donation of a sample for medical research as citizen science is debatable.

Whatever way it is defined, there is one aspect of citizen science that does stand out, namely its decline during the first half of the twentieth century. The precise reasons for this are unknown, but we can speculate that it was associated with the rise of the professional scientist, which in turn was driven by economics.

Chapter 2

The Middle Years - The Twentieth Century

By the early twentieth century the number of citizen scientists who were working independently, funding themselves and pursuing their own particular research interest had declined significantly, and their numbers have remained low to this day. Conversely, the number of citizen scientists who were involved in group projects grew steadily during the twentieth century and increased exponentially around the

FIGURE 2.1 Citizen scientists in action: a botanist volunteer at a Botanical Society of Britain & Ireland (BSBI) field meeting that is aiming to record all plant species found within a square kilometre for the BSBI's *Plant Atlas*.

turn of the twenty-first century. The main reason why the number of independently working citizen scientists declined so rapidly around the turn of the twentieth century and never recovered was the ever-increasing body of scientific knowledge needed, pushed forward by industrial revolutions around the world.

The increasing dominance of professional science

Britain was the first country to undergo an industrial revolution in the 80 or so years spanning the period 1760 to 1840. The boundaries of the revolution cannot be well defined, and industrialisation continued to develop into the twentieth century. In transport, for example, the golden age of canals was between 1760 and 1830, that of railways was between 1840 and 1850 and that of shipbuilding was between 1890 and 1914. The number of cotton mills reached a peak around 1910, and coal mining also peaked in the early decades of the twentieth century. Real GDP per person nearly doubled during the nineteenth century, and by the end of that century Britain was the world's richest and most advanced economy. Such an economy required large numbers of knowledgeable and well-trained scientists and engineers, and a continued steady increase in these numbers. During the nineteenth century there were concerns about the poor standard of scientific and technical education and its inability to meet the demand for trained scientists and engineers. The Great Exhibition of 1851 had simultaneously demonstrated the importance of science and, even at that time, the shortcomings of the scientific education system in Britian compared with some of its counterparts on the continent. For instance, the universities at Cambridge and Oxford had traditionally focused on the arts and treated the sciences as an entertaining diversion for their students.

However, these old established universities were the first to take up the challenge, albeit rather slowly. Degrees in natural science were first awarded by Cambridge University in 1851 and by Oxford University in 1853. The University of London started to award degrees in science in 1860. Later these universities began to run degree courses in which students specialised in a particular branch of natural science. For example, Oxford University added chemistry, physics, physiology, zoology, botany and geology to its list of specialisms in 1886. Numerous colleges and societies which had been established outside London during the nineteenth century formed the basis of new provincial universities, and these were more committed to scientific and technological education. In the 50 years between 1875 and 1925 the number of universities in Britain trebled.

At the same time as there was an economic pressure to provide professionally trained scientists, it was becoming increasingly difficult for amateurs to continue in the same way as they had been doing. As the volume of scientific knowledge increased, the subjects became more specialised and technical, and ever-increasing amounts of time and money were required for amateurs to keep pace with these developments. There were more publications to buy and digest, lectures to attend, learned society subscriptions to pay and items of equipment to purchase. It is calculated that the number of scientific articles published increased at the rate of 5.62% per annum – that is, the number of articles published doubled every 12.6 years.

Working in a professional capacity does to some extent constrain intellectual independence by directing what is taught and how, much to the distaste of many of the old savants. Because of this, earlier in the nineteenth century scientific amateurs would often choose not to become professionals precisely because they preferred to pursue their science as they wished, and they had the funds to do so. One reason why Henry De La Beche turned professional was because his personal fortune had plummeted. He was one of the most prominent gentlemen scientists in geological circles when he was appointed the first director of the government-funded British Geological Survey in 1835. Nonetheless, the growing number of universities and government-funded research institutions that needed experts to fill their teaching posts, as well as the increasing difficulty that citizen scientists were experiencing in pursuing their private research interests, led to a gradual move away from working in an amateur capacity and towards working in a professional one.

Charles Lapworth (1842–1920) is a good example. He was born and raised in Berkshire. His early interests were in the arts, and he trained in Abingdon to be a schoolteacher. However, when he took up his first teaching post in Galashiels in the Southern Uplands of Scotland in 1864, his interest turned to geology. The Southern Uplands have a complex geology, consisting of multiple rock strata that have been variously folded and distorted with the passage of geological time. There are not many fossils there, but Lapworth chose to study the graptolites that are present in large numbers in the dark shale bands around Dob's Linn. Graptolites are marine invertebrates that formed colonies of sometimes thousands of animals which lived in interconnected tubes made of collagen – they resembled small free-floating coral-like fragments. The colony skeletons became fossilised after the animals died and the collagen skeleton dropped to the sea floor. These fossils have very many distinct shapes reflecting the wide range of species, and because of this and the fact that they evolved relatively quickly they are excellent markers of geological time. Confident that the shale bands at Dob's Linn were stratified in strict time sequence, Lapworth was able to associate particular species with rock of particular ages. When he applied this knowledge to the whole of the Southern Uplands, he found that the folding had been so severe that sometimes layers had been overturned, thus becoming duplicated. Unfortunately for the Geological Survey, this meant that their maps were incorrect and the whole area had to be resurveyed.

More significantly still, Lapworth was able to resolve a long-running dispute between two eminent geologists (both of whom were also citizen scientists) concerning the assignment of certain rock beds in North Wales which lay at the boundary between the Cambrian and Silurian periods. Adam Sedgwick believed them to be Cambrian, whereas Roderick Murchison believed them to be Silurian. By careful study of the fossils, Lapworth was able to show that the beds were not associated with either of these periods, but rather they belonged in a new period of their own, the Ordovician, which falls between the Cambrian and Silurian. Lapworth proposed the Ordovician in 1879 when he was still teaching, although it was not given international approval until 1960, 40 years after his death. Nonetheless, Lapworth had become a world authority, and in 1881 he turned professional and was appointed the first Professor of Geology at Masson College, the forerunner of the University of Birmingham, which still retains the Lapworth Geological Museum.

A kind of positive feedback accelerated the move to professionalism and reinforced its grip on scientific research. Once careers based on formal training and accreditation in science provided by universities became generally available, such training and accreditation became a requirement for entry into those careers, and this included the teachers who provided the training. Furthermore, academic research required funding, and the funding bodies needed to be persuaded both that the project merited financial support and that the scientist who was applying for funding was worth supporting. Because the professional scientist required qualifications and plausibility to maintain a living, the amateur became sidelined and rather looked down upon. This extended to the ability to publish the results of scientific work. Although much of the research that was being done would have been virtually impossible without very expensive equipment, some research can be carried out with little more than an enquiring mind and critical faculties. Nonetheless, during the twentieth century it became progressively more difficult to publish results without either a backer who could provide a sufficiently high-profile introduction of one's paper to an editor, or a significantly well-known name on the list of authors.

A good example of this occurred in a short paper published in *Nature* in 1984,[1] in which the two authors described their investigations of the activities of brown hares by very careful and systematic observation. The primary author, A. J. Holley, was not a professional scientist and provided a private address for his contact details in the published paper, whereas the second author, P. J. Greenwood, provided an academic address at the University of Durham. Their findings made an important contribution to delineating animal behaviour, as they described the chasing and boxing activities of brown hares throughout the year. This piece of work was unusual not only because it was citizen science in its clearest form, but also because it was of such significance that it appeared in the premier scientific journal of the day. Such cases are rare indeed, and always have been, which makes it all the more remarkable that this paper was published against a backdrop of professional science. And although it was professional science and the need to make a living as a scientist that for the most part put paid to the sole citizen scientist, computers and particularly the personal computer were going to change the landscape again, so that it would once more favour citizen science.

Latter-day amateurs

For these reasons, the existing amateur scientists were gradually subsumed into the professional system. A few amateur scientists continue to exist – some self-taught, some trained and working in one scientific field but with an avocation in another, some working alone because their ideas clash with the accepted dogma of the

1 Holley, A. J., & Greenwood, P. J. (1984). The myth of the mad March hare. *Nature* 9 (5968), 549–550.

professionals, and some who are so wealthy that they can afford to set up their own research facilities.

Albert Einstein (1879–1955) was a scientist whose hobby was at least as engrossing as his job until 1908, when his fame and genius meant that he could at last swap his job in a Swiss patent office for an academic post. He was a child prodigy whose mathematical intellect soon left his tutors behind; at the age of 12 he discovered an original proof of Pythagoras' theorem. His school years were spent in Munich until the age of 15. Then, in 1894, his parents had to move to Italy because his father's business ran into difficulties. Einstein moved to Switzerland to continue his schooling, and actually gave up his citizenship of the German Kingdom of Württemberg two years later. At the age of 17 he enrolled for a four-year mathematics and physics teaching diploma at the Federal Polytechnic School in Zurich, and after graduating he tried for almost two years to get a teaching post but, remarkably, was unsuccessful, so he took the job of assistant examiner at the Swiss Patent Office in Bern.

Some argue that certain aspects of Einstein's journey outside of academia influenced the formulation of his theories. He spent many hours discussing physics with fellow student Mileva Marić at the Federal Polytechnic. Eventually they became lovers, and were married in 1903. In 1902 he formed a small club to discuss with friends the science and philosophy of contemporary scholars. Even his work at the Patent Office may have helped to shape his thinking on relativity. Some of the patents he examined were related to the thought experiments he conducted on space, time and light. At least Einstein was not hampered by the need for expensive specialist equipment. All he needed for his thought experiments were pen, paper and a quiet space. In 1905, at the age of just 26, this patent examiner rocked the world of physics with the publication of four landmark papers – on Brownian motion, the photoelectric effect, the special theory of relativity and mass–energy equivalence – and the submission of his PhD thesis to Zurich University. Unsurprisingly, Einstein did not continue his work as a patent examiner for much longer.

Hertha Ayrton (1854–1923) was another turn-of-the-century scientist who was a citizen scientist by force of circumstance. She was born into a relatively poor family, and her family became poorer still when her father, an émigré Jewish watchmaker from Poland, died when she was seven years of age. To make ends meet, her mother had to return to work as a seamstress, but fortunately she recognised her daughter's intellectual ability and supported her move to be educated with her cousins in her aunt's school. It was her cousins who introduced Hertha to mathematics and science. In her late teens she became involved with the suffrage movement and formed friendships with several influential feminists, including Barbara Bodichon, a founder of Girton College, Cambridge, and the novelist George Eliot (Mary Evans). She continued to tutor herself once she was living independently as a governess, she passed the Cambridge entry examination with honours and, with financial support and encouragement from her feminist friends, was able to enter Girton College to study mathematics in 1876, aged 22. Four years later she passed the Mathematical Tripos, an elite and very arduous degree that involved a week of back-to-back examinations and was thought unsuitable for women. She was not awarded the degree, as Cambridge would not award degrees to women for another year.

During her four years at Girton, Hertha had begun to demonstrate her penchant for designing and constructing scientific instruments, and when she returned to London, as well as teaching mathematics she continued to develop and then patent her new design for a line divider. This instrument allows a line to be divided into equal parts and can be used for enlarging or reducing images, which makes it very useful for artists, architects and engineers. Again Hertha was given financial help to gain the patents by her feminist friends. Although she would continue to design and patent instruments, her heart was set on a scientific career and to this end she started to attend evening classes at Finsbury City and Guilds Technical College. Here she was taught by, and soon married, Professor William Ayrton. The marriage and a legacy left to Ayrton by Bodichon relieved her of financial pressures, so she was able to pursue her science, which included taking over her husband's work on the properties of electrical arcs. At the time, arc lighting was used in lighthouses and was beginning to replace gas street lighting, but it was very temperamental, hissing and flickering apparently at random. The science behind it was largely unknown, so manufacturers had difficulty overcoming these problems. Hertha Ayrton discovered that the problems were caused by oxygen coming into contact with the carbon rod electrodes, and that removal of oxygen prevented the inconsistencies in arcing, thus allowing her to establish the Ayrton equation, which describes the relationship between arc length, pressure and voltage.

Ayrton published many articles on the science and practical use of electric arcs. As a result she became so well known that in 1899 she was the first woman to be invited by the Institution of Electrical Engineers (IEE) to present her own paper and to be elected a full member of that institution. This was a remarkable achievement considering the conservative nature of the scientific community at that time. The Royal Society, the most prestigious scientific society of the day, refused her membership on the basis that married women had no independent legal status in law. The Royal Society's then President, William Huggins, was also against women 'trivialising' his elite institution.

Ayrton was marginalised by the scientific community and ended up working in a home laboratory. It was only while her husband was alive that she had access to the full facilities of an academic institution. After his death in 1908 her early experiments with what became known as the Ayrton anti-gas fan were conducted in her drawing-room laboratory using the smoke from burning brown paper, and she tested her first working model in the garden of her suffragette friend Ernestine Mill. The fan, which was inspired by her research on the propagation of ripples in sand and water, was designed to clear poison gas from the trenches in the First World War. There was controversy over the fan's efficacy, but by the end of the war 104,000 units of an improved mechanical version were in use, and after the war (and her death) enhanced Ayrton anti-gas fans were used in mines, sewers and other industries that required improved ventilation.

As the dominance of professional science increased through the twentieth century, more lone citizen scientists worked as amateurs for only a small period of their life. Once they had established their credentials in their chosen field, they would often become professionals. Aldo Leopold (1887–1948) and Aubrey de Grey (b. 1963) are two typical examples. Leopold was a keen observer of nature from a young age, and kept journals full of notes and sketches throughout his life. He chose to take a degree in forestry at Yale University because of his interest in nature and the outdoors, and on graduation in 1909 he took up a post with the recently created US Forest Service. The experiences of his early years working for the service, particularly the shooting of wolves and bears, fundamentally changed his thinking about game management and he began to favour a much more ecologically sympathetic approach. His work with the Forest Service allowed him to continue to observe and develop his ideas, and, as he took on more senior roles, to actually influence Forest Service policy. In 1922 he submitted a formal proposal to designate the Gila National Forest in New Mexico as a wilderness area. When it was so designated in 1924 it became the world's first formally recognised wilderness area.

Leopold left the Forest Service in 1928 and spent five years as a private consultant conducting game surveys in the Midwestern states before becoming Professor of Game Management at the University of Wisconsin. This was the first professorship of game management. Thus Leopold had transitioned to working as a professional in the science of wildlife management, a scientific field that he had helped to establish. He had advanced the concept that apex predators bring ecological health, and was fundamental to the reintroduction of bears and wolves into the wilderness areas of the USA.

However, his private studies did not finish there, and their influence on the science of conservation was to be profound. In Wisconsin, Leopold and his family bought a worked out 'sand' farm consisting of 80 acres of land that had been logged, over-grazed and left barren. Together they experimented with what these days would be called rewilding. They planted pine trees and restored prairie on their farm. Leopold documented the changes in flora and fauna as the restoration progressed, and these changes were instrumental in shaping his ideas on conservation and environmental ethics. His sand farm diary and his experiences with wildlife while working in the forest service, together with a formalisation of his land ethics, were brought together in the landmark publication *A Sand County Almanac and Sketches Here and There*. Sadly, in 1948 Leopold died from a heart attack while helping to fight a fire on his neighbour's farm, a year before that book was published in 1949, but it has gone on to have a huge influence on the environmental movement worldwide.

Often the dividing line between the amateur citizen scientist and the professional can become very blurred. When their interest in a particular subject first develops they are clearly amateur, and by the end they are clearly professional, but the point of transition is rather obscure. This is true for Aubrey de Grey. His interest in ageing and the possibility of extending the human lifespan began in his thirties, while he was working professionally as a computer programmer. Today he is President and Chief Science Officer of the California-based Longevity Escape Velocity (LEV) Foundation, a not-for-profit organisation that he founded. En route, he co-founded the Methuselah Foundation in 2003 while still working in software development, and the Strategies for Engineered Negligible Senescence (SENS) Research Foundation in 2009. In the early years he funded the SENS Research Foundation to the tune of US$13 million from his own purse (almost 80% of his inheritance from his mother's estate). All three foundations are researching therapies for countering the processes that de Grey and his colleagues believe cause cell and tissue ageing.

De Grey was born in London, and as a youngster was encouraged by his artist mother to study the subjects in which she was not so strong – science and maths. He took a degree in computer science at Cambridge University and moved seamlessly into a computing career working on artificial intelligence for Sinclair Research Limited, based in Cambridge. Some years later the direction of his work took a sharp turn when he met Adelaide Carpenter, an American geneticist, at a graduate friend's party. She was a trained biologist and visiting professor on sabbatical at Cambridge University's genetics department. They married in 1991. De Grey, who had not had a biology lesson since leaving school, became interested in the subject and taught himself in his spare time by reading scientific texts and being tutored by Carpenter. This put him in the fortuitous position of having the necessary computing and biological knowledge (and contacts) to take over the running of the Genetics Department's FlyBase database, which is a global online repository of genetic and molecular data on the Fruit Fly *Drosophila melanogaster.* He held this job for 14 years, and it gave him the institutional clout necessary for someone without conventional academic training to start a career as a theoretician specialising in the subject of ageing.

Still studying in his own time, he published his first book, *The Mitochondrial Free Radical Theory of Aging*, in 1999. On the title page he is named as Aubrey D.N.J. De Grey, Department of Genetics, University of Cambridge. The book is in fact very well researched and written and cogently argued, and the university awarded him a PhD for it the following year. It summarises the research which indicates that free radicals created when electrons escape the final stages of the process of turning food into cellular energy (in the form of ATP) cause oxidative damage to mitochondrial DNA. De Grey argued that this build-up of mitochondria with damaged DNA is one of the factors that cause ageing, and he proposed solutions that would delay ageing by preventing or repairing this damage. Mainstream academic gerontologists were opposed to this hypothesis for many reasons, including de Grey's arrogance – he seems to have revelled in rubbing them up the wrong way – but mainly because the evidence was correlative and therefore did not constitute strong proof. De Grey points out that it is not a disproof either. His aim always seems to be to get experimental scientists to work on his hypotheses. He has since produced a definitive list of seven causes of ageing, which include mitochondrial DNA damage, and he argues that finding therapies to repair enough of the damage caused by these might allow indefinite longevity, even of people who are alive today, because a significant leap in lifespan, by slowing ageing, would give time for cures to be found for other health challenges. Thus, De Grey argues, we would reach longevity escape velocity. His many conference appearances and the three foundations that he helped to set up all aim to encourage the search for methods of repair. Only time will tell if this search bears fruit.

Our final example of a twentieth-century lone citizen scientist is James Lovelock (1919–2022). He was also largely a theoretician when it came to his major avocation, the Gaia hypothesis, for which he felt that he was unfairly derided by the biologists of mainstream academia. He became an independent scientist and inventor by choice, and was a self-confessed loner who made his living practising science from laboratories he built in the cottages that were his homes, first in Wiltshire and later in Devon. Even with a track record as a professional scientist who would eventually win several prestigious prizes and be elected a Fellow of the Royal Society, he was unsuccessful at times both in winning grants and in publishing papers after he became independent. The first paper that he submitted to *Nature* after starting to work independently was initially rejected with a note from the editor saying 'We don't take papers from private addresses', and Lovelock heard that the editor had remarked afterwards 'they usually come from cranks'.

Lovelock was born in Letchworth, grew up in London, and developed a love of science and gadgets from an early age. He cites a couple of examples – being given a Christmas present by his father consisting of a box of electrical odds and ends when he was four, and selling his stamp collection when he was nine in order to buy the components to make a radio. He hated school but was determined to be a scientist. He could not afford to go straight to university, and instead took a job at a photographic firm while studying chemistry at Birkbeck College in the evenings. Eventually he enrolled for a chemistry degree at the University of Manchester, and although he only graduated with lower second-class honours, he must have impressed his professor, the Nobel Prize winner Alexander Todd, because Todd recommended

FIGURE 2.2 James Lovelock standing next to a 1980s gas chromatograph in his home laboratory at Coombe Mill in Devon in 2011. (Dave Wilkinson)

him for a post at the prestigious Medical Research Council laboratories at Mill Hill in London. Here he used his scientific knowledge to invent and construct novel instrumentation, and he was given the freedom to pursue ideas that intrigued him. Indeed, he says of his most famous invention, the exquisitely sensitive electron capture detector (ECD), 'I had no reason to invent it'.

The ECD would be his passport to becoming an independent scientist. After around 20 years at Mill Hill he developed a hankering to work independently. He was given the opportunity to do so when NASA, aware of his ECD expertise, invited him to consult with them on the creation of sensitive instruments for the study of planetary atmospheres in order to look for extraterrestrial life. It was while working on this at the Jet Propulsion Laboratory in Pasadena, California that Lovelock had a Eureka moment and came up with his Gaia hypothesis. He proposed to NASA that the best way to search for extraterrestrial life was to study the entropy of the atmosphere. For example, the Earth overall has low entropy. The high concentrations of oxygen and small amounts of methane in the Earth's atmosphere make it chemically unstable, so something must be maintaining this highly reactive yet low-entropy state. Lovelock was interested in the basis of this paradox. At the same time he learned that although the Sun has been getting hotter over geological time, the temperature on Earth has remained remarkably stable. Lovelock's Gaia hypothesis posits that life on Earth interacts with its physical surroundings, the rocks and the atmosphere in a synergistic way to self-regulate and maintain a benign environment for life. To say that the hypothesis was contentious when it was first put forward is an understatement. Many biologists hated it, believing it to be teleological and in conflict with the theory of natural selection. What would be the mechanism by which the world's organisms worked together to maintain the environment? However, Lovelock was convinced, and for at least the first ten years after proposing the theory he devoted as much of his time as his paid scientific work would allow to developing the Gaia hypothesis further. Therefore, during this phase at least, Lovelock was working on his hypothesis as a citizen scientist.

Both Lovelock and De Grey, whether they were right or wrong, stimulated a huge amount of scientific research around the world, which has advanced our understanding in many disciplines in a way that might otherwise have taken years. It is more usual nowadays to refer to these highly trained lone scientists as independent scientists – indeed that is how Lovelock described himself – while using the term citizen scientist to refer to people who take part in group science projects.

The growth of group citizen science

Group science took off quite gradually in the twentieth century in parallel with the increase in disposable income, leisure time and good schools, and the falling cost of books and other publications. By 1900, access to school education had greatly improved, and advances in printing and distribution had significantly reduced the cost of books and magazines, but the average salary was only just over £40 per annum, and many people had to work six days a week with little holiday. A clerk might earn a guinea (£1.05 in decimal money) a week, but even if he was living frugally (being single, not drinking alcohol and eating only vegetarian meals) his basic cost of living would be just over 19 shillings (£0.95 in decimal money). Professionals fared better. A teacher might earn £60 to £100 per annum, and the lowest-grade commissioned officer in the army might earn £200. Their cost of living would be higher but they would have some disposable income. For instance, a professional earning £700 per annum could have a disposable income of £200. Thus, even though group citizen science tends to be mostly observational in nature and therefore low in cost, the early contributors were mostly drawn from the middle or upper classes. Three of the earliest group projects, all of which are still continuing today, involve recording phenological events, recording plant distribution and bird ringing.

Recording phenological events costs very little. Almost everyone would have been able to recognise the event – for example, the date of the first snowdrop flower, first apple blossom or first frogspawn – without the use of a field guide. However, it is only group science if recording is done in a systematic way, with everybody recording the same events and the information being gathered centrally. Nevertheless, contemporary phenologists are extremely grateful for the data they have been able to extract from the few surviving diaries of individuals, such as those of Robert Marsham and his family (see page 13), which help to fill in the gaps between periods of systematic recording. The latter requires central planning, organisation and collation of the data, which is usually undertaken by a society or club.

The Royal Meteorological Society (RMetS) organised the first countrywide collection of phenology records in the UK, even though collection of phenological data had started many years before the society existed. It was the Belgian astronomer, meteorologist and mathematician, Adolphe Quetelet, who appealed for phenology to be recorded as part of national statistics, when he was attending a British Association meeting in 1841. The British Association formed a committee and took four years to come up with a huge set of events to monitor, extending the phenology to a large number of plants, fish, mammals and birds. Discussions rambled on and it was not until 1875, 30 years after Quetelet's appeal, that the RMetS adopted a more practicable version of the scheme. The society, which went through a couple of name changes during this period, was by then 25 years old and had about 400 members worldwide. The still significant cost of membership, at £2, would have excluded many interested amateurs. Nevertheless, the majority of the members provided private rather than academic addresses in Britain – that is, they were citizen scientists. Membership continued to grow throughout the twentieth century, reaching a peak of 3,691 in 1997. The number of individuals taking part in the phenology recording scheme peaked at around 600. Unfortunately, the RMetS abandoned the project in 1948, when members questioned why a meteorological society should be interested in collecting information about the timing of biological events.

Even though it was curtailed prematurely, the almost 75 years of recording by the RMetS provided important baseline data on species' responses to temperature before the advent of global warming. It also demonstrated the synchrony that exists between species that have an ecological dependency. A nice example of this synchrony is that between the date of emergence of the Orange-tip butterfly *Anthocharis cardamines* and the flowering of Hedge Garlic *Alliaria petiolata*. Orange-tip caterpillars feed on the seed pods of two wildflowers, Hedge Garlic and Lady's-smock *Cardamine pratensis*. Consequently, it is important that when the adult butterflies emerge these plants have had enough time to flower and develop seed pods to the right stage for the caterpillars to eat. The butterfly's emergence and the growth and flowering of Hedge Garlic and Lady's-smock are both affected by the temperature in early spring. Over the 75 years during which the RMetS collected data for the first flight of the Orange-tip and first flowering of Hedge Garlic, their records of mean March temperature were in the range of 2–9°C, although in most years the temperature was in the range of 3.5–7.5°C. The date of flowering ranged from 9 April to 8 May, and the date of emergence ranged from 20 April to 20 May. Plotting the date of emergence and the date of flowering against the mean March temperature shows that both occur earlier as the temperature increases. The rate of change is similar across the 3.5–7.5°C temperature range, but at higher temperature there are indications that they diverge, with the emergence speeding up and the flowering slowing down, which may be a long-term problem for the Orange-tip.

Between 1948, when the RMetS scheme was abandoned, and 1998, when a new scheme was begun, phenological recording was once again an individual activity for those with an interest in the timing of natural events in their locality. Although a large amount of data has certainly been lost, the records of two citizen scientists stand out because their diaries have survived and indeed been published. Jean Combes, a naturalist from Ashtead in Surrey, started collecting the date of budburst for a single oak tree in 1950, and continued to record the date for the same oak tree until she was well into her nineties, with the help of her family. Although from year to year there is wide variation in the timing of budburst for this tree, over the total recording period of around 70 years there is a clear trend, with budburst happening three weeks earlier by the end of that period. These data have been used by government departments and scientists interested in the impact of climate change, and in 2009 Jean Combes was awarded an OBE for her services to phenology. She commented that 'I don't feel that I have done very much really, except living a long time'.

The timing of leafing in oak is important for the many associated species that are affected by it. For example, many of the spring flowers of the woodland floor, such as Bluebell *Hyacinthoides non-scripta*, Wood Anemone *Anemone nemorosa* and Lesser Celandine *Ficaria verna*, fit their growing season into the period between the end of winter and mid- to late spring, their leaves already withering as those of the tree canopy are expanding. Similarly, hatching moth caterpillars rely on the timing of oak leaf expansion, as they need to eat young leaves as these have low levels of tannin. Blue Tit in turn relies on the timing of the flush of caterpillars to feed its chicks. If the time available for the growth of the spring flowers is reduced, or the timing of the tits' egg laying gets out of synchrony, the entire ecosystem is harmed.

Richard Fitter (1913–2005) attended the London School of Economics and subsequently worked for non-governmental planning organisations, and, during the Second War World, in Operational Research for Coastal Command. He was obviously a keen amateur naturalist because he spent his evenings studying natural history, and published his first book, *London's Natural History*, just as the war ended. He went on to make his living from writing and from working with a number of nature

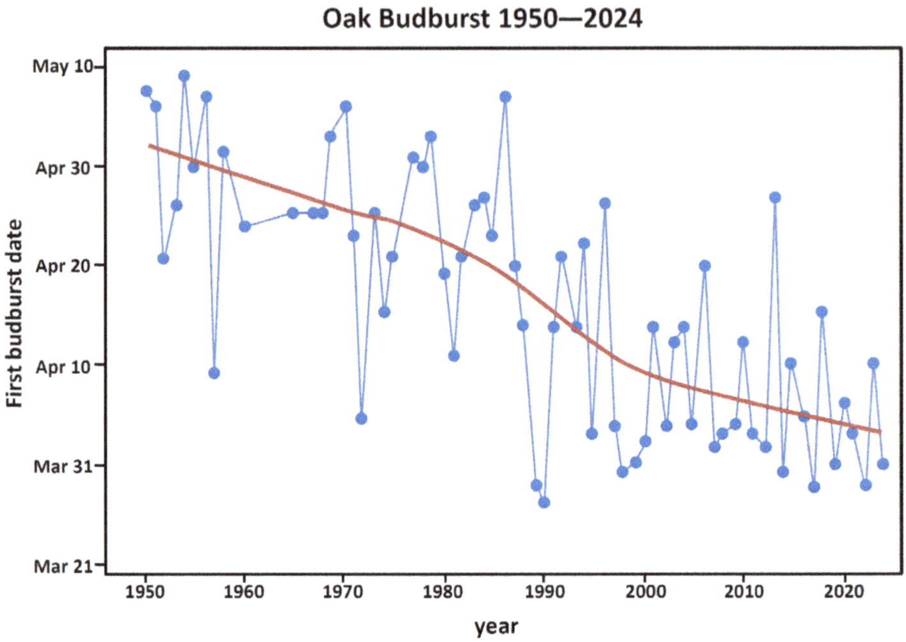

FIGURE 2.3 A graph showing Jean Combes' data on the annual date of budburst on a single oak tree. The trend for earlier budburst over the recording period is clearly visible. (Woodland Trust)

conservation organisations. In the 1950s he began to make observations of over 500 wildflowers, and, importantly for phenology, the first flowering date of 385 of these species. After 47 years of recording, and in collaboration with his son, a professional academic ecologist, in 2002 he published an analysis of the flowering data in the journal *Science* (Fitter & Fitter 2002). His data showed that the average first flowering date was 4.5 days earlier in the 1990s compared with the previous four decades, revealing 'the strongest biological signal yet of climate change'. In total, 16% of the 385 species actually flowered significantly earlier, an average 15 days earlier, and 3% flowered significantly later. The analysis suggests that global warming could affect the structure of plant communities as a result of this variation in response. This has become a landmark paper in the scientific literature on climate change, and by 2023 it had been cited 265 times in other publications. Like Jean Combes, Richard Fitter seems to have been fairly modest about his achievements. He appeared on the BBC television programme 'Springwatch' in 2005, surrounded by his notebook diaries, and is reported to have 'seemed amused that the data within them had proved to be so important'.

His paper in *Science* was not Richard Fitter's only contribution to citizen science. His later published works included some of Britain's earliest true field guides – portable books with the contents arranged specifically to aid species identification. In 1952 he published *Collins Pocket Guide to British Birds* (Fitter 1952), which was the first field guide to British birds, and it made him a bestselling author. *The Pocket Guide to Wild Flowers*, co-authored with David McClintock (McClintock & Fitter 1956), followed in 1956. He wrote many more books with various co-authors, the last one being *The Wild Flowers of Britain and Ireland* (Blamey, Fitter & Fitter 2003) which was published in 2003 when he was 90, and was co-authored with his son Alistair and the artist Marjorie Blamey, who was 85. His botanical field guides together with those by many other authors were an essential resource for the increasing numbers of citizens who were helping to map the distribution of wildflowers in the second half of the twentieth century.

Interest in the geographical distribution of plants had begun to develop in the first half of the nineteenth century, with the publication of *Outlines of the Geographical Distribution of British Plants* by Hewett Watson in 1832 (Watson 1832). Watson (1804–1881) was what the Victorians called a philosophical naturalist – someone who focused on explaining nature rather than simply observing it – and he would today be described as a citizen scientist. His absorption with botany probably started when his mother sent him to work with the family's gardener to take his mind off family problems. After starting training for the legal profession in Liverpool, he then moved to study medicine and natural history at Edinburgh University. There he met Robert Graham, another citizen scientist who was interested in botany and was training to become a doctor. It was Graham who encouraged Watson to investigate plant distribution, and Watson actually privately printed *Outlines of the Geographical Distribution of British Plants* just after he left Edinburgh. He then worked professionally as a phrenologist for about six years, but he had inherited an estate in Derbyshire in 1833, and this gave him the financial freedom to eventually focus on botany and his other main interest, evolutionary theory. Between 1847 and 1850 he published the four volumes of *Cybele Britannica: or British Plants and their Geographical Relations* (Watson 1847–1850), and in 1873–1874 he published the two volumes of *Topographical Botany* (Watson 1873–1874). The latter might be described as the first atlas of British plants in that it lists all of the known locations of the flora at the vice-county level. The vice counties are perhaps Watson's most lasting legacy. They are largely based on the old county boundaries, and are still in use today in the organisation of botanical recording.

Watson's fledgling curiosity about plants was part of a growing amateur interest in botany outside of universities. The first two botanical societies in Britain had their inaugural meetings in 1836. The Botanical Society of London (BSL) was formed, by a group that called themselves the Practical Botanists, at a meeting in the Crown and Anchor Tavern on the Strand, with the objective of 'the advancement of Botanical Science in general' and especially species descriptions, the reading of papers and the formation of a library. The Botanical Society of Edinburgh had similar aims but was more closely aligned with the University's Botany Department. By 1838 the BSL had about 75 members resident in Britain, a large proportion of whom were amateurs, each paying between a half and one guinea per annum. Even so, the society was by then already stretched for cash, and appealed in the 1838 Proceedings for members to buy a copy of the Proceedings. Before long Watson took over the leadership of the BSL, streamlined processes and, among other things, encouraged members to pursue the factors that determine the distribution of species. Many of the members exchanged plant specimens or contributed specimens to herbarium collections, and these 'vouchered' specimens would have helped Watson to compile his distribution data. Although there was ongoing interest among amateurs, and the society continued to grow, in 1856 it was dissolved due to serious financial problems.

Out of the ashes, exchange clubs were formed by members of the BSL who wanted to continue to swap specimens. These eventually crystallised into two main clubs – first the Botanical Exchange Club, which was joined by many former members of the BSL, and then in 1884 the Watson Botanical Exchange Club, which had a predominantly new membership consisting of comparative beginners. A little earlier, in 1878, another society with a slightly different remit was founded. This was the Botanical Locality Recorder Club. It did not collect and exchange specimens but rather searched for and collated new localities for plant species. These were published in its first quinquennial report for 1877, at county level using Watson's vice counties, and could thus be regarded as addenda to Watson's *Topographical Botany*, expanding the distribution data within the latter. The report was very sensitive to the opprobrium the club had received from sections of the established botanical community which suggested that the club was 'a kind of co-operative society for the repetition of already published plant stations' and that its members were nothing more than 'conscience-less grubbers-up of rarities'. It seemed especially keen to reassure Watson about the quality of its data and the validity of its work. Watson was known to be prickly and ill-tempered in his exchanges with colleagues, and in *Topographical Botany* had 'expressed the belief that there are hardly fifty botanists who sufficiently comprehend the philosophy of plant-distribution to take any living interest in a series of data'.

It is true that some members of these clubs and societies collected rare plants. The most notorious such collector was George Druce (1850–1932), a pharmacist by profession but also an avid amateur botanist. He is one of the few people to have written more than one county flora. In fact he published four – for Oxfordshire, Berkshire, Buckinghamshire and West Ross, between 1887 and 1929. In 1903 he became the new and dynamic Honorary General Secretary of a flagging Botanical Exchange Club, and the changes that he made brought in many new members, but he was a controversial figure and alienated some people. On one occasion, when taken to see a field of orchids, he 'pulled up specimens by the armful' and a fellow field botanist had to be restrained from thumping him. Despite the controversies and travails, the more extensive recording of plant distribution carried out by the members of these clubs and societies between 1880 and the start of the First World War invites comparison with contemporary distribution data for the first time.

The next significant development for amateur botanists who were contributing to the study of plant biogeography was at a conference of the Botanical Society of the British Isles, now renamed the Botanical Society of Britain & Ireland (BSBI), in 1950, at which a project to map the distribution of British plants was proposed. The BSBI was a reinvigorated Botanical Exchange Club whose membership had plateaued at just under 500. With the launch of the scheme, membership soared to around 1,400 by 1962, when the first atlas was published, helped no doubt by the general increase in leisure time, wealth and interest in the environment. However,

THE BOTANICAL SOCIETY AND EXCHANGE CLUB OF THE BRITISH ISLES.

(VOL. IV. PART II).

REPORT FOR 1914

OF THE

BOTANICAL EXCHANGE CLUB

(Conveniently Abbreviated for Citation REP. B.E.C.)

BY THE

EDITOR AND DISTRIBUTOR,

R. H. CORSTORPHINE, B.Sc.

The Subscription, 7s 6d per annum, and Non-Contributing Members' Subscription of 5s per annum, become due on January 1, 1916, and should be sent to

G. CLARIDGE DRUCE,
YARDLEY LODGE,
9 CRICK ROAD, OXFORD.

Cheques for three or four years in advance would save much trouble and expense in postage.

Parcels for 1915 should be sent post paid, on or before 10th December 1915, to A. R. HORWOOD, Esq., Municipal Museum, Leicester.

The Distributors and Editors for 1916 will be W. H. PEARSALL, Esq., and D. LUMB, Esq., Dalton in Furness.

Printed by T. Buncle & Co., Arbroath.

November 1915.

FIGURE 2.4 The title page of the 1914 issue of the journal of the Botanical Society and Exchange Club of the British Isles. This was in George Druce's heyday, when as Honorary General Secretary he re-invigorated the Botanical Exchange Club. Note his keenness to collect subscriptions for years in advance. (Botanical Society of Britain & Ireland)

the Maps Scheme, as it was called, was to be so comprehensive, including all native species and many aliens mapped on to every 10 × 10 km square (hectad) in Britain and Ireland, that it relied on many more volunteers than just the BSBI members. This was one of the first projects run by professional scientists that relied on the mass involvement of members of the public – the very first one involved watching for migratory birds. The project was innovative in many ways. The plant atlas was to present dot distribution maps aligned to the Ordnance Survey grid, with a dot for each hectad where the species was present. Recorders in the field were provided with special record cards that listed abbreviated plant names and numbers. These could be quickly summarised on a hectad master card. A special card that listed just the 113 commonest species, named in full, was used by many schools. Finally, new mechanical devices developed specifically for the project meant that map production was swift, and made it possible for the BSBI to achieve its planned publication date. The devices worked with punched cards and allowed an average map containing 1,000 dots to be printed in 20 minutes.

The whole project was a great success and spawned mapping projects for many more taxa, ranging from fleas and spiders to birds and butterflies. It had an immediate impact on conservation, too, with the publication in 1977 of the first British Red Data book, in which rare species were defined as occurring in 15 or fewer hectads, and their declines were shown by comparing pre- and post-1960 hectad totals. However, academics made little use of the data, probably because of the difficulty of handling such large data sets without access to computers. The popularity of field surveys with the volunteers continued to grow, as did the number of records collected and the number of species covered. This is reflected in the growing number of BSBI members, which today stands at about 4,000. Two more atlases have been produced, one in 2002 and one in 2022. The atlas that was published in 2022, *Plant Atlas 2020*, covered 3,445 species using 26 million individual species records collected over 178,000 volunteer recording days. Despite the problems of recorder bias and recording consistency due to so many people being involved (an issue to which we shall return in Chapter 6, and another reason for the slow take-up of the data by academics), the 70 years of comprehensive, countrywide recording demonstrate clearly what had been known instinctively, but had until then been unproven, namely that native plants are suffering. The ranges of more than 50% of all native plant species and ancient introductions have shrunk, mostly due to changing land use and the way land is managed. In Britain, one in every five species is now listed as threatened.

The last of our three group citizen science projects that burgeoned in the twentieth century required very much less data collection, but the collection relied on chance and the data were few and far between. The appearance and disappearance in spring and autumn of some of Britain's birds used to be a great mystery to naturalists. Aristotle had hypothesised that when swallows disappear in the autumn it is because they have gone into hibernation in crevices. Then in the sixteenth century the Swedish priest Olaus Magnus suggested that in fact they hibernate in mud at the bottom of lakes and ponds, a notion that was still believed in Gilbert White's time at the end of the eighteenth century. Other theories also existed. In the 1680s, an English minister,

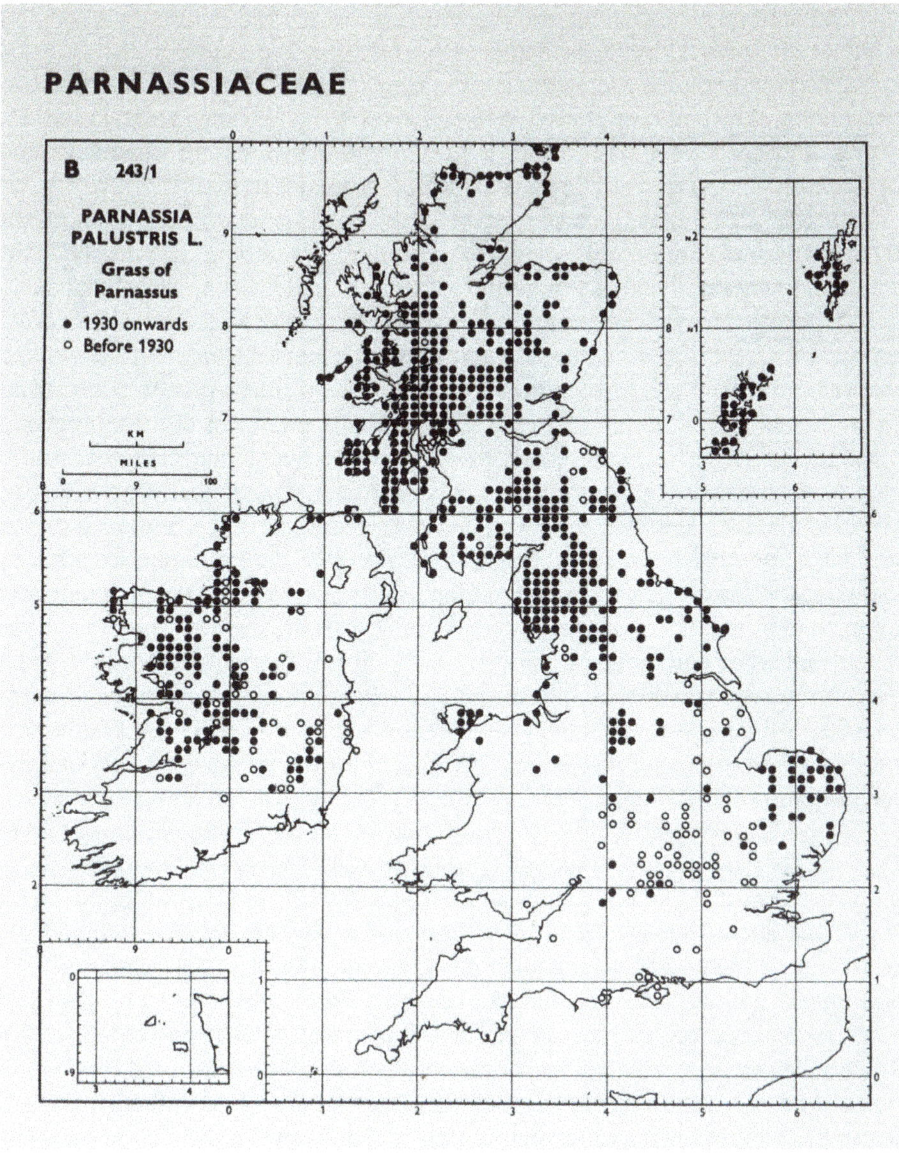

FIGURE 2.5 An example of the innovative dot distribution maps used in the BSBI's first plant atlas published in 1962. This map is for Grass of Parnassus, *Parnassia palustris*. A retreat northward is clearly highlighted by the map. It is due to the loss of its lowland habitat through land drainage, not climate change - it happened much too early for it to be that. (Botanical Society of Britain and Ireland)

Charles Morton, suggested that birds migrate to the Moon for winter. By the end of the nineteenth century sufficient clues had been observed to lead ornithologists to believe that the appearance and disappearance of the birds was due to migration, and attempts were made to mark birds to try to prove this. It was a Danish school-teacher, Hans Mortensen, who developed and was the first to use aluminium rings, stamped with an identification number and address, in 1899. A number of ringing schemes using Mortensen's technique were started in Europe and North America at the beginning of the twentieth century. The first was in Germany in 1903. Two bird-ringing schemes were started in Britain in 1909 by two keen amateur birders, Arthur Landsborough Thomson and Harry Forbes Witherby. A third scheme was run by *Country Life* magazine, but it failed almost as soon as it started.

Witherby (1873–1943) worked for the family publishing firm from when he left school until his retirement in 1936. His avocation was birding, to which he was devoted, spending all his spare time and vacations studying ornithology, and taking his family with him. His even taught his new wife how to skin birds while they were on honeymoon! In 1907 he founded and edited the journal *British Birds*, one of the main aims of which was to encourage cooperative work among ornithologists, and when he founded his ringing scheme it was under the auspices of the journal and called the British Birds scheme. Like Witherby, Landsborough Thomson (1890–1977) was an avid amateur ornithologist. He studied natural history at Aberdeen University, served in the First World War and then took up a career as a medical researcher. He set up his scheme while he was in his first year at university. This well-organised scheme, known as the Aberdeen University Bird Migration Inquiry, involved both ringing and trapping of birds, but it closed during the First World War. However, the British Birds scheme persisted and indeed expanded rapidly. In 1937, when Witherby handed it over to the British Trust for Ornithology (BTO), it was ringing almost 50,000 birds annually. Witherby was an early benefactor and keen supporter of the BTO, which runs the ringing scheme to this day. The BTO set up a committee to oversee the scheme, and Landsborough Thomson was its first chairman, serving until 1965.

There are conflicting reports as to which species was the first British ringed bird to be found abroad. One source suggests that it was a Scottish Lapwing found in France, another claims that it was a Cumbrian Common Tern found in Spain, but in fact it seems that the honour should be given to a Eurasian Wigeon, which was ringed as a duckling in Sutherland under the Aberdeen University Bird Migration Inquiry in 1909 and caught later the same year in a duck decoy in the Netherlands. The mystery of swallow migration began to be solved with the first recovery of a ringed bird in South Africa in December 1912. The identification of South Africa as the swallow's winter home came as quite a surprise to ornithologists at the time, and Witherby wrote in *British Birds* that it 'seems to me extraordinary'. For migration studies, most of the recaptures occur by chance when the bird dies or is accidentally trapped in some way. There are some bizarre examples, such as gulls and ducks being hit by golf balls, a Reed Warbler caught by African spiders and an Osprey eaten by crocodiles, but many recaptures are from wildfowl being shot by hunters. The records gradually built up as the number of ringed birds increased, and in 1926 Landsborough Thomson was able to write a book, *Problems of Bird Migration*, which described the routes, directions, distances and other factors involved in bird migration.

All of this effort relied heavily on volunteer ringers and, rather like amateur botany in the years leading up to the mid-twentieth century, the ringing effort struggled financially, with some of its income coming from the ringers themselves. In the early 1950s the BTO started to receive funding from the Nature Conservancy, whose rationale for providing funds was the realisation that in order to do its scientific work it needed the help of amateur naturalists to gather the information. Again, as in amateur botany, there was a sharp increase in the number of amateur ringers in the late 1940s, which was reflected in the annual number of birds ringed. One could almost use the same graph to show the relative number of botanists or the number of birds ringed plotted against year in the twentieth century – just swap the vertical axis as required! By its hundredth year, in 2009, the BTO scheme had ringed 36 million birds, and today close to one million birds are ringed in Britain every year by 2,600 trained ringers, most of whom are amateur volunteers.

We now know a great deal about bird migration – the Swallow flies to South Africa in the winter, the Sand Martin migrates to Morocco, the Manx Shearwater flies to southern Argentina and the Arctic Tern migrates to Antarctica – but there are still many species whose journeys are unknown. In 2002 this knowledge was finally published in the first migration atlas. This could not have happened any earlier because of the need for computers to perform the comprehensive and powerful analysis required. By that time computers and computer networks had not only become essential for analysis of the big data sets that were being collected, but ringers were also able to keep their records on their own computers and submit them electronically to the central database, just as plant recorders were able to do with the records collected for *Plant Atlas 2020*. The people who find bird rings are now also able to report them electronically. In fact three landmark events in the world of computing in the late twentieth century had a profound effect on big-data citizen science, namely the launch of the Internet and the Windows PC in the mid-1980s, and the public availability of the World Wide Web from 1993 onward.

The development of fast, cheap computers made it possible to handle huge amounts of data, far more than could ever be processed by hand. This has enabled novel methods of data handling to be used, but certainly in the early years it did not herald a renaissance in statistical techniques. Larger data sets and faster computers gave the impression of data being of greater value, but what it really meant was more data giving the same result, but with an extra significant figure. In broad terms, the analysis did not improve very much by using ever larger amounts of data, it just became more reliant on computers to carry out the mathematics. This is especially so in biological systems where unpredictable events happen. Statistical methods of analysis that were originally created in the nineteenth century and the first half of the twentieth century were designed for small samples, with calculations done by hand, which would be extrapolated to a population. In the case of computerised citizen science it was not necessary to limit the numbers or even regard the data as a sample in the traditional sense.

Chapter 3

Big Science Now

During the last quarter of the twentieth century and the early twenty-first century the personal computer became widely available, and this piece of equipment enabled individuals to handle scientific data in the same way that professionals could. It was the speed of computation that computers could achieve which set them apart from log books and slide rules, and the development of engineering and production techniques allowed ever smaller computer chips to be produced. Moore's law (originally formulated in 1965 and then revised in 1975) is a reflection of this technological change. Gordon Moore postulated that there would be a doubling of the number of components in integrated circuits every two years. This is an extrapolation from observations, and although it is known as a law it is in fact merely a long-standing projection. What has happened along the way is that the apparent computational speed has increased and the amount of data that can be dealt with

FIGURE 3.1 Citizen scientists in action: entomologists looking for butterflies. (Matthieu Gauvain/CC-BY-SA-4.0)

has grown in proportion to the capacity of computers. This ability of computers to handle huge amounts of data has given rise to the interesting phenomenon of high-volume citizen science. What we mean by this is that for many projects the emphasis has shifted towards the collection of vast amounts of data that can be arranged and organised by computers, and away from citizen science as a thinking and observing activity.

Fast, cheap computers have certainly made it possible to handle far larger amounts of data than can be processed by hand. This has enabled novel methods of data handling to be used, but in the early years it did not herald a renaissance in statistical techniques. Larger data sets and faster computers have given the impression that data are of greater value, but what is really happening is that more data are giving the same result, but with an extra significant figure. Statistical analysis methods that were originally developed in the nineteenth century and the first half of the twentieth century were designed for small samples with calculations done by hand, which would then be extrapolated to a population. In the case of computerised citizen science it was not necessary to limit the numbers or even to regard the data as a sample in the traditional sense.

Data-driven projects

Within very large, data-driven projects we can see three broad categories where the scientific content has been displaced from the individual to a group. The first category consists of the centrally started projects that originate in the same way as any scientific investigation, but rapidly run out of steam due to lack of large-scale statistical data. In cases such as these, attaching the epithet 'citizen science' enables the project coordinators to encourage and recruit potentially vast numbers of willing volunteers. This will allow enormous amounts of data to be generated that may or may not be useful for clarifying the questions that are being asked by the project. It is this sort of project that is found in healthcare systems where mass screenings are being carried out. The advantage to the individual of being included in the project may be obvious – for example, in the case of bowel cancer screening, as the earlier such cancer is detected the better the outcome for the patient. However, for the overall project, unless the samples are examined for other biochemical markers, the figures for the rates of detection and survival among those individuals in whom bowel cancer is detected will not alter. The point here is that having more data does not necessarily lead to having more information. In the case of bowel cancer screening it may be that rates of detection will alter over time, reflecting a change in prevalence. This in itself would provide ample evidence for the need to look at external factors that could be affecting the change, such as diet and exercise.

The second category of large-scale projects consists of those that are started and controlled by a group of concerned or interested individuals. These sorts of projects generate considerable involvement among the general population, as they are usually associated with an area of common concern. This is most often associated with environmental problems, such as water pollution of lakes, rivers or sea shores. Although in some parts of the world air pollution is of great concern, this does not tend to generate the same public involvement as water pollution. Projects such as these can either generate their own data, or use already available data that are in the public domain. Whatever way they are obtained, these data are used to put legal pressure on those responsible for the clean-up that should inevitably follow pollution that has been demonstrated by such detailed analysis.

The third category of citizen science project arises from the desire to monitor a constantly changing situation when this can only be achieved by large-scale public involvement. Projects of this sort include bird migration studies and population counts of insects, most often butterflies. We shall discuss the importance of these projects later, but for now suffice it to say that they have already provided an enormous amount of information to balance against the weather patterns associated with global warming.

We live in what we are told is a data-driven century, but it is only data driven because we can use computers to handle the numbers. In past centuries, when such tools were not available, the methods used to handle data and extract information were relatively simple statistical techniques that used a population sample from which extrapolations could be made to the total population. In the twenty-first century the focus has changed to nearly always being based on the acquisition of data, as if this is sufficient justification in itself. Sometimes large-scale data collection is necessary to develop systems and models that can aid our understanding of natural phenomena, as in weather forecasting. However, when conducting citizen science projects in biology we need to be careful that the amount of data does not swamp the underlying picture and the ability to understand what is being observed.

FIGURE 3.2 Collection of water samples for analysis of chemical pollutants requires no more equipment than you would need if you are looking for plankton in pond water.

It is also possible in this philosophical explanation of how science works for data to produce a paradigm shift by creating a simpler explanation for a particular set of circumstances. This was exactly what happened with the Copernican Revolution in cosmology, when there was an abrupt move from the geocentric ideas of Ptolemy to the heliocentric explanations of Copernicus for the movements of the planets and stars. It is said that the shift was made not because the Copernican system included better and more accurate predictions of celestial events, but because it was easier to use and gave simpler mathematical solutions to the observed movements. This, by implication, meant that it was sufficiently flexible to be developed further in the future. So bear this in mind – your project may have unforeseen results and be of use further down the line.

Bringing large-scale citizen science projects back from just generating theoretical implications to providing understandable information for the citizens involved is paramount. The best way of doing this is by communicating results in as understandable format as possible. One of the primary and most important activities for all sorts of citizen science projects is to keep those involved as well informed as possible. To this end the European Citizen Science Association (ECSA) has produced a ten-point list of principles that projects should aim to incorporate from the planning stage onward. Most of the suggested principles are straightforward and self-evident, helping to define what citizen science is. Some of the statements are easy to understand – for example, 'Citizen science projects actively involve citizens in scientific endeavour that generates new knowledge or understanding' and 'Citizen science projects have a genuine science outcome'. Some of the other statements are much more general and are concerned less with the scientific component than with the social implications of projects. This is hardly surprising, as citizen science has itself become a legitimate topic for sociological study. It is partly due to this interest that it has also become a political tool for integration and inclusiveness studies, as demonstrated by the Cooperation in Science and Technology (COST) programme, which encourages women, young researchers and those from under-represented countries to participate in projects. This works well with large projects that span geological and political boundaries, but is not so effective for small local groups that are working unsupported and for a specific reason – for instance to study a polluted river.

A very good example of citizen science working both for the scientific community and for the citizens who have actively collected the data can be seen in a Swedish study of radioactivity across the country. The study was prompted by fallout from nuclear weapons tests and then by the more recent Chernobyl nuclear reactor disaster in Ukraine in 1986. Although the study was started many years later, in 2018, in relation to the half-life of radioactive waste it is chronologically still very relevant. The project was conducted on a grand scale, with secondary schools being invited to participate in the process of collecting data. Eventually the number of participants was whittled down to 135 secondary school classes with ages in the range of 13–16 years. They collected samples of fungal fruiting bodies, along with associated soil samples, but also looked out for wild animal droppings that could be collected, identified and tested.

What made this a true citizen science project was that the schools weighed and identified the fungal fruiting body samples before sending them to various departments in academic institutions to allow detailed radioactivity measurements to be carried out. The project also had a very specific aim, which would hopefully have a very practical outcome, namely to determine the residual radioactivity in the sporocarps (the caps) of fungi. This was deemed important because they are widely collected and sold in Sweden. The two radioactive isotopes that were of interest were caesium-134 and caesium-137. The number of each of these two isotopes is the atomic mass – that is, the combined number of protons and neutrons in the nucleus. As the number of protons remains constant within an element, the difference in atomic mass is solely due to a change in the number of neutrons present. This apparently small difference makes a big difference to the isotopes – caesium-134 has a half-life of about two years, whereas caesium-137 has a half-life of just over 30 years. The half-life refers to the time it takes for the quantity of a radioactive substance to decrease by half. In practical terms this means that the radiation from caesium-134 will take about four years to decrease to a quarter of its original level, whereas caesium-137 will still be highly radioactive when the caesium-134 has almost ceased to be a problem. By the time the project started, caesium-134 was regarded as of little importance, whereas caesium-137 was still present at about half the level detected in 1986.

The situation is complicated by the nature of the biological systems in which caesium is found. Living organisms use energy to take up chemical compounds from the soil in which they grow, and to maintain a specific chemical composition within their cells. As a result, unwanted elements can sometimes become concentrated where they are least wanted, and where their presence will stop the normal functioning of the cell, which is what happens with caesium when it enters the food chain. This highly reactive element does not normally cause disruptions in plant cells, but does accumulate as an unwanted salt. In animal cells, however, as it is in the same chemical group as lithium and sodium it becomes widely distributed throughout the body. It was this biological accumulation of radioactive caesium that was going to be investigated in edible mushrooms.

Once the schools had responded to the invitation to join the project they were supplied with a research box containing tools and instructions for foraging as well as a simple radioactivity monitor. It was recognised from the very start that a clear manual was needed for the teachers to use, as well as details of how they could access support if they had any questions. Documentation and the availability of expert help have both been consistently shown to be crucial for the smooth operation of citizen science projects. Although the box contained a simple radiation detector that could be calibrated in a basic way, its main purpose was to indicate the relative differences in radioactivity levels between two samples. The project started in late 2018 and continued into 2019, with schools sending their samples on to testing laboratories for precise evaluation.

When the results were collated, the distribution of caesium-137 was as expected, with the same distribution as had originally been mapped after the Chernobyl incident in 1986. The results also showed where the caesium levels still exceeded the permitted safe levels for the sale of mushrooms. Once the data had been collected, the scientists from Uppsala University and the Swedish University of Agricultural Sciences, who had organised the project, helped the participants to understand and make use of the results.

Understanding radiation and its hazards is quite challenging, due to the fact that different kinds of radiation vary in their effects on different biological tissues.

This has led to the development of several different methods of expressing units of radiation. In terms of the energy of radiation, the Gray (Gy) corresponds to 1 joule/kg, or 100 rad (radiation absorbed dose). The rem (roentgen equivalent man) takes into account the biological response, and the Sievert is the effective dose and is the unit normally quoted for exposure limits. In the case of X-rays and gamma rays the unit of exposure is the roentgen. All of these units refer to the amount of energy that is trapped by the biological material, which is difficult to comprehend in a practical situation such as consumption of wild mushrooms. Although it may seem that there are already more than enough units to deal with, the Swedish academics introduced a new unit designed specifically for their subject matter, namely the mushroom sandwich. When overlaid on a map of Sweden, the number of mushroom sandwiches in a pile at any point represented how many sandwiches, each containing 200 g of mushrooms, one would need to consume daily each year in order to receive an additional 1 mSv of radiation from the caesium. This allowed a very easily understood dosage map to be imaginatively created for a very specific use. To put that figure of an additional 1mSv into perspective, in the UK we receive about 2.6 mSv a year from a variety of sources as a result of natural exposure to radiation.

In an additional questionnaire the participants were asked how they had benefited from the project. Overall there seemed to be an increased awareness of radiation and its measurement. The number of responses to the follow-up questions, as well as the provision of data by the schools, demonstrate that with an adequate level of interaction not only do participants provide information, but also they learn more about the subject they are studying. Although there have been legitimate questions about the scientific content of many citizen science projects, the Swedish example demonstrates how the scientific flow is bidirectional, with participants on both sides of the project learning something new. However, for many other projects it is not so easy to see how the supply of data by citizens will lead to mutual exchange of knowledge.

Flow of information in ornithology

The question of direction of information flow in large-scale projects has to be continuously addressed because it is linked to two potential problems. The first is that it is assumed, often incorrectly, that citizen science projects will improve the control and governance of knowledge by individuals and groups. The second potential problem, which is rarely considered, is that giving third parties access to the accumulated data can lead to their misuse. This second problem particularly relates to large, powerful digital companies being given access to data, much of which may be personal. Such companies are transnational by their very nature, and because they are commercial businesses they are not accountable to anyone in particular. Therefore they can use citizen science data, especially medical information, for their own purposes. They do certainly carry out some brilliant science, but they also keep a tight hold on any results that may be of commercial value, which implies that they would have little interest in democratising science that might reduce their profit. This clash between open and closed and between private and public science is much more pronounced in areas of basic research than in applied areas such as environmental and conservation science, where individuals with little or no background knowledge can participate in data collection. It is for this reason that citizen science is rarely encountered in basic science studies. In these areas a great deal of prior knowledge may be required to even formulate a question in such a way that it can be answered. Nonetheless, involvement in citizen science projects has definitely fostered scientific literacy, as exemplified by the Swedish study described earlier, where the organisers went so far as to look at how involvement had changed the participants' perceptions of radiation and how it is measured.

Another way in which citizen science projects can involve participants is by having them collate observations or results that have already been collected, sometimes many years previously, as in the case of phenology studies. Better still, citizen scientists can look for interesting trends using data that would have been collected as a by-product of another project, or even as casual observations associated with a hobby. This was an area tapped into by Wells Woodbridge Cooke (1858–1916) in 1881. As we mentioned earlier, applied biology is a particularly fertile arena in which citizen science can work, and bird migration is one of the most suitable areas of all. To track not just the arrival of migratory species but also the routes by which they travel requires people along the migratory passages to make a note of when birds pass through their area. What Cooke managed to do was to move from making his own observations on the timing of birds' arrival and departure to involving large numbers of individuals who would then supply him with information to be collated. He started his observations in 1881, and over the years these developed and became known as the North American Bird Phenology Program, which generated huge amounts of information that was written down by ornithologists on Migration Observer Cards. This massive repository of data was accumulated from 1881 to 1970, and has now once again become a citizen science project as the cards are digitised by volunteers for future use under the auspices of the USA National Phenology Network.

When Cooke first started his project he was working as a teacher, and ornithology was his hobby. Interestingly, it is said that he showed an early interest in natural history, and at the age of 12 he was given a rather unusual present with which to pursue his hobby – a gun for shooting birds to collect as specimens. He graduated from the University of Iowa in 1879 and became a teacher in the Indian Schools system. It was while he was teaching that he became interested in the migration of birds. He had already started to make his own notes on the passage of birds when in the winter of 1881–1882 he asked the ornithologists of Iowa to send him information about the bird species that were winter residents, as well as the dates when migrant species first arrived. The network of observers grew to such an extent that he was able to collate data for migrant species across the whole Mississippi migration route, now called the Mississippi Flyway, which runs up the centre of the USA from the Gulf of Mexico to Canada. It was not long after this study had started that the American Ornithologists' Union (AOU) took over the coordination and development of the project. The continued development of the observations and increasing scale of the study meant that by the late 1880s it had become too large even for the AOU to look after, and it was passed on to the Division of Economic Ornithology. This was later renamed the Bureau of Biological Survey, which subsequently morphed into what is now the US Fish and Wildlife Service. When it was taken over by the Division of Economic Ornithology there were 3,000 volunteers covering virtually the entire USA and Canada, with part of the West Indies represented as well. When it ceased to function in 1970, the massive archive of individual observation cards was passed to the US Geological Survey.

Cooke's work is particularly interesting in the context of citizen science because not only did he motivate large numbers of individuals to generate usable data, but also he was himself a citizen scientist – he was conducting much of his ground-breaking work as an ornithologist while employed as a teacher. This is a model that has been reused many times in many different applications with citizens making and recording their observations. The participants are often described as non-expert, but this does them a disservice. In ornithological studies where an expert needs to be able to clearly and accurately identify species, unpaid and formally untutored bird watchers can be described as expert as they are often able to identify birds from their calls or from their movements and flight patterns, as well as from their plumage. It is also true that when different levels of skill are involved, if incorrect identifications are made in big projects involving large numbers of individuals, the results will not be significantly affected by this. Therefore, for some of the largest projects, the more people who become involved the better the result, and encouraging people to join in regardless of skill level will improve the overall picture. A global project called eBird, run by the Cornell Laboratory of Ornithology, allows users of its system to send photographs, sound recordings and sightings of birds to a shared database that almost by coincidence now shows clear maps of migrations and population changes when overlaid on satellite maps. The fact that it is truly international has resulted in a claimed 100 million sightings annually, and the app is now available in more than 30 different languages.

The RSPB and the British Trust for Ornithology (BTO) also run schemes for non-scientists who, by participating, gain an insight into the changes in number and distribution of bird species in the UK. Started in 1979 via a children's television programme the RSPB scheme is called the Big Garden Birdwatch, whilst the BTO scheme, launched in 1995, is called, perhaps a little confusingly, Garden BirdWatch. The BTO scheme was designed solely to collect data on birds, but has become so popular and useful that it now generates data about how wildlife as a whole uses our gardens. Originally, BTO members were asked for weekly lists of birds seen in their gardens. In 2003, the survey was extended to include mammals, reptiles, amphibians, butterflies and bumble bees. Dragonflies were added to this list in 2011. In addition to simple lists of those species seen, considerable detail is included, such as garden features and any food put out for visitors, as well as any sick or dead specimens. The latter has become Garden Wildlife Health as a collaboration between RSPB, BTO the Zoological Society of London and Froglife. Both schemes are carefully designed to help people to connect with wildlife and learn observation skills.

As an archive of wildlife the Garden BirdWatch holds a significant amount of informative data, as well as some amusing accounts, such as the report of a tawny owl taking up residence in an old Christmas tree as a daytime roost. In surveys that involve large numbers of participants, such as the Garden BirdWatch and the Big Butterfly Count, errors of identification that would skew the results of smaller surveys will become statistically non-significant, while the overall data create a rolling image of changes in population density and species distribution. By generating data and maps that participants can browse it also returns information to the original data collectors, and enhances the value to all concerned. This ensures that the participants feel they are playing an active part in the project and that it is a two-way process.

Just as the BTO decided to instigate a citizen science project to monitor bird numbers and distribution, so too did the RSPB. The version started by the RSPB differs in detail but broadly reflects the same picture of bird numbers as the BTO survey. The RSPB have called their survey the Big Garden Birdwatch and claim that it is the largest citizen science wildlife survey in the UK. This may or may not be so, but with a claimed number of participants of more than 500,000 each year, it is a significant indicator of the general public's attitude to birds, and presumably also to other wildlife, in the UK. The most significant aspect of this survey is that it takes place every year on the last weekend of January – having a specific date helps to give a snapshot of numbers. Each participant registers and then picks an hour to record what they have seen. In this respect it is very similar to the Big Butterfly Count organised by Butterfly Conservation. The citizen scientist should only record birds that have landed, not those that are just flying over the garden. They also record, where possible, the numbers of birds. This can be quite difficult, as small birds such as the Long-tailed Tit often travel in quite large groups and never stay still for long enough to be accurately counted. It was quickly realised by the RSPB that 'no birds' was also an important result and this, along with the positive counts, has alerted us to the declining numbers of birds in our gardens. Furthermore, it has highlighted the rises and falls in population numbers and changes in feeding behaviour of wild birds. The BTO, the RSPB and the Joint Nature Conservancy Committee have developed a spin-off, which is a Breeding Bird Survey that monitors bird breeding in an area by regular visits by individuals to a site over the spring and summer.

Citizen science has also been used to provide data for a project on bird distribution and breeding in Turkey. Starting out as a project for ornithologists and then more generally involving birdwatchers and birdwatching clubs, this survey has been adopted by the Turkish Ministry of Agriculture and Forestry. The large-scale nature of the survey has resulted in a great deal of information about both avian biodiversity and important breeding sites in Turkey. The large-scale mobilisation of citizen scientists has been particularly important because the area of Turkey is more than three times that of the UK.

Flow of information in entomology

In the countries of the northern hemisphere there are a number of biological groups that have relatively small and distinct numbers of species, the identification of which can be easily handled by amateur scientists. Three such groups are birds, with 632 species known in the UK, butterflies, of which there are 57 UK species, and orchids, with about 57 UK species. Such relatively small numbers of species within any one group make it easier for the amateur to learn enough to make a quick and accurate identification. It also means that even though some individuals may not regard themselves as ornithologists or even as birdwatchers, they can join in surveys of bird distribution by noting down those species that they are certain they can identify.

Another important aspect of large-scale projects that should be addressed in the planning stage is to make it easy for participants to both record and return their results. In the UK there have been many large-scale projects, the running of which requires the investment of considerable time and effort on the part of the organisers to make sure that every participant feels that they are a genuine part of the scientific endeavour. There are some very good examples of how this has yielded excellent results in the UK. Butterfly Conservation runs the Big Butterfly Count, a very easy project in which to participate. All that is needed is for participants to pick a site between the dates of the survey, which are usually in July and August. The surveyor then sits for just 15 minutes and records all of the butterfly species that they see. Butterfly Conservation supplies a downloadable chart to aid identification, as some groups, such as the blues and skippers, can be a bit tricky. The results are then added to the database. Butterfly Conservation uses these data to produce interactive maps of species distribution and the commonest species seen overall. In areas where there are few participants it is even possible to see one's own contribution to the project. This is an excellent example of using the data to give participants a clear picture of what the local species are and in what sort of numbers they occur.

Across the whole of Europe it has been demonstrated that most citizen science projects involve making and recording observations, or simply collecting data, rather than experimental work or developing research questions. There are of course exceptions to this, as we have seen in the Swedish project on radioactivity in mushrooms, and as happens in many river-monitoring projects where training is given in the accurate and correct use of equipment. Monitoring projects, especially those directly associated with biodiversity studies, benefit from citizen science involvement as it increases the scope of these projects far beyond what could be achieved solely by short-term paid research based on fixed-length grants. This is exactly how the BTO and RSPB managed to extend their initial project to encompass all manner of other animal groups. Indeed, it is claimed that more than 80% of biodiversity data is collected by citizen scientists. There are obvious caveats to this, as data collection to study the geographical distribution of plants and animals depends on the inclination and ability of collectors to travel to the appropriate site of study. Gaps in the distribution of recorders can result in geographically lopsided data. This has been seen in the past – for example, in Victorian times the distributions of various species of beetle tended to reflect nineteenth-century holiday destinations of

FIGURE 3.3 The type of information available from the Big Butterfly Count to help participants to gain maximum benefit from the project.

the middle classes rather than true distributions. This is most apparent with obscure groups, for which there were often only a handful of collectors, rather than groups such as butterflies for which there were large numbers of hobby collectors.

Among all the surveys associated with citizen science, the most common biological subjects are birds. This is hardly surprising, as they are the most obvious and easily targeted group of species for identification and counting. The only other comparable group would be the Lepidoptera, or more specifically butterflies. However, in areas beyond northern Europe and the UK the sheer number of butterfly species can become quite problematic, particularly in families where the butterflies tend to be small, such as the Lycaenidae (the blues, coppers and hairstreaks) and the Hesperiidae (the skippers). Although many other butterfly families also become more speciose the further south one goes, the blues, like the skippers, tend to be small and fast, which makes identification quite tricky when the species are all broadly of the same shape and colour.

Flow of information in botany

It has long been realised that heavy metals in the environment are harmful to plants and animals, and this is amply demonstrated by the paucity of plant life on spoil heaps from mines where high concentrations of these elements are found. It should be remembered that some of these metals are only toxic in large amounts, and are in fact necessary for good health, such as iron, cobalt and zinc. Others, such as mercury, lead, cadmium and arsenic, are highly toxic even at very low levels. Although there is no scientific definition of a heavy metal, they account for 90% or more of the known elements, depending on the definition you use. So using a broad phrase like 'heavy metal' can be very useful in setting up a project as long as there is a clear understanding of which metal is being measured. Using the idea of heavy metals in a broad sense, the Austrian Science Fund has supported an interesting and informative survey based in and around Vienna. This survey is called the Heavy Metal City-Zen, and was designed to investigate heavy metal pollution in soils and crop plants in the urban gardens of Vienna.

The participants in Heavy Metal City-Zen were asked to grow specific crops of spinach and radishes in pots filled with soil from the garden and, for comparison, in pots of commercially produced compost. It did not matter if there were differences in growth, as commercial compost is designed to encourage growth, so these plants would be expected to be bigger. After a specified time period the plants and compost were sent to the organising laboratory, where they were tested. The compost was tested for its original heavy metal composition, and the plants were tested for the quantity of heavy metals taken up. It is, of course, essential to measure the original levels of heavy metals, as even if uptake is passive rather than active, a higher concentration in the plants would be expected if there is a higher concentration in the soil. The project focused on measuring lead, cadmium and zinc levels. Most samples were well below the acceptable safe limits, and a few were above the acceptable limits. The project examined not just the concentrations of heavy metals, but also their mobility and bioavailability, and whether they could move into the food chain. It is mobility and bioavailability that is important, as metals in the soil are not intrinsically hazardous – only when they enter the food chain do they become a problem. The project extended over a complete growing season, from April/May 2020 to 2021, and had the additional complication of having to work around the significant restrictions that were imposed from 2020 in response to the COVID-19 pandemic. This is a rare example of a citizen science project in which the participants were involved in conducting an experiment (with the careful separation of plants into experimental and control groups, and the plants then being labelled and sent off), rather than simply making observations.

Birdsong and migration of birds and mammals

Unlike the Viennese project on heavy metals, some citizen science projects require long-term funding and dedicated participants. One such project was a very imaginative wildlife study of the geographical delineation of Yellowhammer dialects in the Czech Republic. The Yellowhammer *Emberiza citrinella* is a small passerine bunting with a very distinctive song. It has long been recognised (from work published in 1869) that yellowhammers have singing dialects. Once the project had been started in 2011 and the first distribution maps produced from the data, there was an upsurge in public interest. So great was the interest shown by birdwatchers and ornithologists that the project was extended.

During the first year of the yellowhammer dialect project enough data were collected to distinguish between two major dialect groups. The volunteers recorded the easily recognised song and then sent it to the collating laboratory at Charles University, Prague, where the recordings were mapped to give a visual representation of the data. In the first year most of the volunteers came from the Czech Ornithological Society, but the idea of there being dialects among birds caught the wider public's imagination to such an extent that newspaper articles and radio programmes on the subject resulted in a growing pool of active volunteers. The organisers also realised that they could actively use such radio broadcasts to recruit volunteers in specific geographical areas where records were scarce. This was important because there was an unbalanced distribution of recordings. Where there were few residents or few visitors, the number of recordings was also much reduced. As an increasing amount of data became available, another problem arose relating to the quality of the recordings. Most were of good quality, but an increasing number had to be discarded because of technical issues such as inaudibility and lack of clarity. Further investigation revealed that this was due to the recording devices not having an adequate bandwidth for the vocal range of the song. To overcome this problem, volunteers were sent a test sound to verify the effectiveness of their own recording equipment.

To further motivate the participants and keep them interested in the project, rather than simply having one-directional data traffic towards the scientists running the project, a game was created. The aim of this was to improve the overall geographical coverage of the recordings by basing it on geocaching or orienteering, where specific locations are searched for using accurate map locations or GPS data. Interestingly, introduction of the game doubled the geographical range of data within 12 months.

Over a five-year period, although most of the administration and development of the project was overseen by two PhD students working part-time, seven scientists and 166 volunteers were creating a database of recordings. This stalwart group managed to achieve about 90% coverage of the Czech Republic with approximately 4,000 recordings of yellowhammers singing.

The project in the Czech Republic was so successful that it inspired similar citizen science projects in several other countries and eventually an international project to compare dialects among yellowhammer populations across the native ranges of the UK and then the introduced population in New Zealand. Later on it encompassed many other European countries, increasing the overall understanding of bird populations and their birdsong. It also aided the categorisation of the distribution of bird populations and motivated people to go outside and experience nature. This project, like so many large-scale studies, could not have been carried out without citizen scientists. No matter how well planned and organised such projects are, they need a large number of individuals to collect the data.

It is not just birdsong projects but also any large-scale observations that require feet on the ground, and it is these citizen science projects that have clarified much of what we know about bird migration, and the only way for the science to make sense is for such studies to be international. A very obvious example of this would be studies of bird migration. The large number of volunteers that these studies manage to recruit indicates just how much affinity the general public has for nature in general and bird life in particular.

It has been suggested that the initiation of many projects coincides with the tacit acknowledgement of the important role of citizen science in answering questions relating to agriculture, as ever growing populations need to be fed, and local knowledge and experience are invaluable for optimising crop production. The involvement of citizen science in agriculture has a long history, perhaps best demonstrated by 1,910 years of recordings of locust outbreaks in China. These data have now demonstrated that their value extends far beyond local use, as they have shown a distinct link between locust population explosions and climate fluctuations.

Citizen science in agriculture has a special role in cultivating a bidirectional transfer of knowledge. The reasons for this are clear. It is only the goodwill of the farmer which ensures that accurate information is passed to the scientist about, say, a pest, so it is incumbent on the scientist to pass back to the farmer all of the information that he has discovered, to help to ameliorate the problems caused by the pest under investigation. This apparently idyllic situation contrasts with most projects, where data may travel in both directions but rarely impinge on the livelihood of the citizen scientists.

In 2002 a citizen science study was established specifically to look at conflicts between farmers and wildlife in Namibia. This project epitomised the bidirectional flow of information, knowledge and, above all, understanding of a problem. The study was looking at the conflict between the wild Cheetah *Acinonyx jubatus* and the ability of farmers to maintain their livestock. The cheetahs would on occasion regard the livestock as dinner, hence the very real conflict. Starting in 2002, the project was originally a collaboration between the Namibian authorities, the farmers and a wildlife research group from Leibnitz, Germany. The importance of the farming community playing a central part in the study was recognised at the design stage, and it quickly became apparent that some degree of trust between the farmers and the scientists was essential, particularly as conservationists sometimes tend to view the loss of domestic animals to predation only from the perspective of the predators.

Although the project was primarily interested in the relationship between cheetahs and farmers, it was recognised that Leopard *Panthera pardus* was also an occasional predator of the farmers' livestock. This conflict of interest between the farming community and the conservation of these big cat species had led to farmers killing quite large numbers of the predators in order to protect their herds. Big cats tend to range widely, which means that they often cross international boundaries. Also, because they are obligate carnivores, they tend to need a lot of space in which to hunt. Although the figures vary widely, it has been estimated that cheetahs occupy land at a density of about one individual per 70–100 km^2. There are estimated to be around 1,300 cheetahs in Namibia.

As a primary aim of the project was to try to resolve the problem of loss of domestic livestock without culling the cheetah population, it was essential to integrate the research into the farming community. Therefore three research stations were set up in the farmland being studied, in order to foster trust in the scientists among the farmers, and patience while waiting for the outcome. The cheetahs were tracked, and scats, droppings and hair samples were collected by both farmers and scientists. By making the farmers primary participants in the project, the amount of information and knowledge that could be passed on to the collective group was increased. For instance, the farmers knew how to trap the live cheetahs, and this information was passed on to the scientists. The box traps that the farmers constructed were important because, unlike most big cats, cheetahs do not feed on carrion and so cannot easily be lured to bait – in fact they generally do not feed on anything they haven't caught themselves. The cats were trapped so that a blood sample could be taken and a GPS tracker fitted. This allowed information to be obtained on whether a particular individual had a preferred prey species and a preferred range.

It was discovered that cheetahs are opportunists and that they regularly take domestic livestock, but none of the tracked animals did so selectively. The farmers gained knowledge of the movements and feeding of cheetahs which enabled them to manage the livestock on their land and remove them from areas where they might attract cheetahs. This had an immediate effect in reducing livestock losses to predation by cheetahs. As a result of the success of this citizen science project looking at cheetahs a similar project was set up for leopards.

It was the involvement of farmers as primary participants in the team that was vital for the project's success. This also generated long-lasting trust among all the team members, including the farmers and the scientists from both nations involved. A large part of the land on which the cheetah project was carried out was shared in common.

However, when large areas of land are enclosed and are owned by individuals, as in Europe, it can be difficult to run extensive biodiversity studies without the explicit cooperation of the farmers and landowners. Of course it is best if the farmers join in because they are interested and motivated to do so, rather than feeling pressured to take part. It is for these reasons that comprehensive biodiversity studies are relatively rare in European states. On the other hand, biodiversity projects that are carried out on private land such as gardens can be highly successful, but can give a skewed overall picture of species distribution and population size simply because gardens tend to be highly managed and often contain few larval food plants.

What all large-scale citizen science projects in biology have in common is that they generate data that would otherwise be impossible, very difficult or prohibitively expensive to obtain. Successful citizen science projects often generate results that appear in peer-reviewed scientific journals, and this has been happening increasingly often as projects become more professionally run. Although recognition through publication is a relatively new phenomenon for citizen science, historically there is no doubt that unnamed citizen scientists generated much of the data that was published. They would be observing, and sometimes recording, data about the natural world, but always passing on information about the plants and animals that they encountered. This would include knowledge of animal migration and hibernation. The accuracy would often be compromised by gaps – for example the question of where migratory birds went in winter – which in the past would have been filled by folklore that was created and developed to explain the unknown. For instance, as mentioned earlier in this book, long before attempts were made to document the passage of migratory birds, suggestions were made as to where they spent the winter. These theories attempted to explain how these very delicate fliers could apparently disappear in autumn and return in spring, especially as they seemed far too small and fragile to travel long distances.

Biodiversity

As we have already intimated, studies of biodiversity on farms are intrinsically difficult because most farmland is privately owned. Consequently, citizen science projects that are designed to count and collate species on farmland need the explicit help and assistance of the farmers themselves. It is often assumed that knowledge of the land has been passed down through the generations to the current farmer, and that they therefore know best. However, the discontinuity caused by changes in agriculture throughout the twentieth century shifted agriculture from stewardship of the land to growing crops regardless of the effects, with the result that the common knowledge of land care and agriculture was largely lost. Intensification of farming during and after the Second World War led to the end of traditional low-yield cultivation of fields. This kind of one-way change is a usual result of changing agricultural practices – for example, the alteration in land ownership caused the open field system that had been in operation for around 1,000 years in the UK to change from community ownership to private ownership. Alongside this alteration in ownership came a cessation of ridge-and-furrow ploughing and a considerable increase in available pasture, much of which moved to meadows, which we now consider traditional.

These changes over the course of history are reflected in present-day changes, with industrialisation of agriculture moving knowledge of the land away from the farming community and towards the academic and citizen groups that have a vested interest in studying and hopefully maintaining biodiversity.

In Austria, to resolve this problem of lack of interest in the non-commercial aspects of agriculture, an initially educational project – *Biodiversitätsmonitoring mit Landwirtinnen* – was launched. This was sufficiently successful to gain its own momentum and turn into a biodiversity project that acknowledges the important role of farmers in the preservation and promotion of biodiversity. Austria is unusual in having a very large area of retained alpine meadows of different sizes. There are around 8,400 meadows, with about 20% of agriculture being based on alpine farming. The flora of the pastures that are used for summer grazing varies depending on the types of animals for which they are used. This is also true of lowland meadows, which like their alpine counterparts can have a huge diversity of plants and animals.

At the other end of the scale there is an area of scientific endeavour, previously confined to technical laboratories and academic communities, but now a legitimate part of citizen science, which is associated with the handling and manipulation of DNA. This aspect of citizen science was very popular at the end of the twentieth century, but there now seem to be diminishing numbers of citizen scientists involved in it. This kind of work has implications for both ethics and safety, and there are many regulations controlling the kind of experimental research that can be carried out on DNA and organisms in established laboratories. However, the information necessary for anyone to cultivate microorganisms or extract DNA is freely available (Wall 1999). Although there are a number of these so-called DIY biologists looking at various genomes, their numbers seem to be diminishing, probably because considerable technical expertise is required for most types of genetic manipulation that go beyond simple extraction of the genetic material. Although it is not impossible to develop the necessary technical expertise in a domestic kitchen, getting negative results can be frustrating for the citizen scientist who is trying to develop these skills in such 'kitchen labs'.

In Western countries there are significant legal controls that make it difficult for individuals to buy the specialist equipment and reagents required, even for some of the most basic genetic manipulations. There are also ethical codes of safety. For these reasons, concerns about this form of citizen science are occasionally expressed in the press, but in general they are unfounded. Although there is potential for misuse, this has rarely been an issue.

In contrast to this type of research, most citizen science projects are not dependent on a particular skill or on prior knowledge of the subject. Nevertheless the outcomes are almost invariably positive both for the individuals involved and for the public and scientific communities. For the individual, the benefits may be as simple as learning new techniques and gaining knowledge that is not easily accessible in a classroom setting. For many projects there is an advantage to wider society, as this work engenders a greater sense of belonging to a social group that cares about the environment.

Data and analysis errors

A perennial question associated with all manner of citizen science projects is the validity of the data that have been generated. There are many different types of data, ranging from simple identification of species to chemical analysis of river pollution. It is important to remember that questions about validity are not confined to citizen science, but apply to all good science. In the case of citizen science it is important to be aware of potential pitfalls relating to the type of data and the manner in which data are collected – as with so many things, the planning stage is vitally important. In small groups, such as a team of scientists, a change in methods can be executed with great speed if necessary. With large groups of citizen scientists, which may number several thousand, any change can be very difficult to implement.

We might have a broad idea of what is meant by accuracy of data, but it is worth considering what we mean by accuracy and precision, and the difference between the two. Precision is a measure of how close together values are – for instance, on a dartboard it is how close together the darts land, regardless of where they are or whether they are anywhere near the specified target. In contrast, accuracy reflects how close the various data points are to the true value – on a dartboard it is how close the darts are to the target value. However, a couple of points about the accuracy of results should be borne in mind. First, a wide scatter of data points can still be regarded as accurate if their mean value coincides with the true value. Secondly, the true value is not usually known, so it is here that statistical techniques are used. Associated with the two measures, precision and accuracy, are two types of error, systematic and statistical. Systematic errors affect precision and statistical errors affect accuracy.

Systematic errors consistently either over- or underestimate a true value, which in the context of a project where observers are identifying species would result in consistent misidentification of one species as another. The equivalent statistical error would be the misidentification of a single species as a number of different species. Both types of errors, but particularly systematic ones, can be difficult to unravel without further investigation. Such misidentifications are best dealt with by checking dubious results, but this puts considerable strain on the identity checkers. With regard to the statistical errors, because they are random if large enough samples are collected they will be lost in the overall picture and consequently can be ignored. This approach to handling random misidentification is all very well, but often accuracy is important in order to avoid rare species – which tend to be those that are misidentified most often due to lack of familiarity – being lost from surveys altogether. Even more misleading, rare species can be misidentified as a similar common species because of a bias caused by the identifier assuming that a rare species would not be found.

When setting up citizen science projects, a primary consideration should always be how to set a standard for the data being collected. Although it is possible to alleviate and on occasion eradicate statistical errors, this does make data handling very complicated. In the case of identification data, without systems in place it is almost impossible not to end up with some misleading information. This is a very important point, as citizen science is pivotal in creating distribution and population maps and species density data.

Much has been written about a perceived gap between academic science and citizen science. However, in a real citizen science community where there are professional scientists at work as well, trust between the groups is not only paramount for success but will ensure accuracy of the outcome.

This is not to say that errors will not arise, as it is almost axiomatic that they will, but they can be dealt with by many different methods. Even during such seemingly straightforward processes as transforming data from the written records of the Migration Observer Cards to a digital form, errors can be created. These are largely excluded by checking the transcription against the original, but that will not pick up errors in the original record. Therefore we have to accept that some data will be inaccurate, but in citizen science projects that is perfectly acceptable. As you can see, one of the most effective techniques for avoiding errors is checking, and although this does create more work within a project, it is well worth it for the increased robustness of the results.

There have been surprisingly few tests of data integrity in citizen science. Integrity here refers to both precision and accuracy – what we could describe overall as data reliability. Looking at the range of opinions on this in biological citizen science projects, two points seem to stand out. First, if it is a direct measurement of something like chemical content, there may be greater variation between results obtained by untrained participants than between those made by professional scientists who are more dexterous with the equipment. This may in part reflect group size, as the larger the group the more likely it is that outliers (odd data points) will arise. With individuals who have had only rudimentary training, as would be expected with citizen science projects, the analyst will be working to their own standards and consequently although their results may well be consistent within themselves they may be less consistent between similar individuals. The second point is associated with identification recording schemes – the results obtained by citizen scientists and professionals would be expected to be very close together when the species that are being identified are unambiguous. This situation is most often found in bird counts, where distinguishing species can be learned quite quickly. At the other end of the scale, identification of many insect groups, or more broadly invertebrates, can be very difficult, requiring specialist knowledge and sometimes even a microscope. Nonetheless, the motivation and developed skill of people who would otherwise be termed amateurs should not be underestimated.

The best way of dealing with the almost inevitable anomalies and inaccuracies of identification is to have verifiers who can check records that have been presented for inclusion in a project. This has been done with great success in citizen science studies of birds and butterflies, and particularly with studies of another insect group that has been studied for its diversity and numbers, namely bumble bees.

Bumble bees, data flow and errors of identification

We can all recognise bumble bees, and as there are only about 24 or 25 species resident in the UK it would be easy to assume that they should be easy to identify. However, this is not the case. Many bumble bee species are very similar in appearance, they can be quite variable in colour and the colours can vary with the time of year. They also have caste systems, so workers and queens can differ significantly, as can the males. This difficulty of identification is not confined to bumble bees, but is widely found in recording projects involving invertebrates. A large proportion of the early projects in which amateur recorders were involved had recruited a self-selecting group of individuals who were highly competent, with well-developed expertise in identifying their chosen group. By the time that large-scale projects started, inevitably these included recorders who had more enthusiasm than skill. This skills gap increases with groups such as the bumble bee projects, where identification can be challenging.

Bumblebee monitoring projects attract a large number of enthusiastic people who may not have a particular interest in bumble bees per se but are concerned about the reported decline in pollinators and therefore want to be involved in these sorts of monitoring projects. All of the bumble bees in the UK belong to one genus, *Bombus*, so for a project to work effectively it is important for identification to be made at species level. This is in contrast to other invertebrate identification studies, where very often identification down to genus level is all that is required. Even so, projects can lean heavily on expert verification of records by viewing the photographs that accompany each supplied record. This does lead to a higher number of records being discarded due to inadequate photographs. It is this data quality that can dog citizen science projects and can disillusion participants if their data is constantly questioned due to its lack of clarity. Perhaps surprisingly, little research has been carried out on the magnitude of this problem, even though it would help other projects to gain a sense of what needs to be done to maintain a consistent standard of data across a disparate group with wide variation in levels of competence.

In 2019, one of the rare comparisons of projects was made and published by Steven Falk. This analysis looked at two citizen science projects – Blooms for Bees, organised by Coventry University, and Bee Watch, run by Aberdeen University and the Bumblebee Conservation Trust. As both of these projects were identifying bees to species level it was reasonable to compare the success rate for each of them.

The Blooms for Bees project ran during 2017 and had responses from 485 citizen scientists who were keen to be involved. These volunteers supplied 4,200 observations over the year. In contrast, Bee Watch provided records from 2011 to 2015 and involved a total of 3,427 observers. As would be expected, this much larger group produced many more observations – there were 11,509 records in total. It was realised that it was important to check the identification of bees, so the volunteers were asked to provide photographs of the bees that they recorded. For both of these projects there was basic training given online, supported by feedback given to the individual recorders. Checking the records was a slow process, and the Blooms for Bees project suggested that it would take about an hour to check 35 records. In contrast, the Bee Watch project reported that about 240 records were checked per hour. This discrepancy was in part attributed to Bee Watch participants being encouraged to take photographs with digital cameras rather than smartphones (which had a lower resolution).

After checking, 72% of the Blooms for Bees results and 82% of the Bee Watch results were accepted. In total, 20% of the Blooms for Bees records were not verifiable and 8% were not bumble bees. In contrast, 11% of the Bee Watch records were not verifiable and 7% were not bumble bees. The number of species confirmed was quite different for the two studies, with 15 species for Blooms for Bees and 22 species for Bee Watch. This disparity was expected in view of the length of time for which Bee Watch ran.

In both projects, practice over time improved the performance of individual recorders. However, accuracy tended to be low compared with that in other insect studies, such as Ladybird Watch or population studies of butterflies. One of the interesting findings to emerge was that over a prolonged period the overall accuracy within the project did not improve. This can be explained by the fact that the constant turnover of individuals, with practised and reliable observers leaving and new observers joining, means that the overall standard remains the same.

Bees and butterflies are highly regarded by the general public, and therefore it is easy to recruit people to citizen science projects involving these insect groups, but one group that has more of a problem engendering affection is the wasps. Nonetheless, it was decided that a large-scale citizen science investigation into the genetic diversity of wasps would be of value, and so the Big Wasp Survey was developed. The aim was to investigate the nature of isolated populations gradually drifting away from the founding genetic form into isolated populations and varieties. This project inevitably involved a lot of citizen scientists working across the country. It was coordinated by University College London, the Natural History Museum, the Institute of Zoology, the University of Cambridge and the University of Gloucestershire. It was also partly funded by the Royal Entomological Society.

This was always going to be a tricky project because instead of the participants being asked to spot wasps so that simple distribution and population data could be gathered, the group of scientists taking part wanted to involve the citizen scientists in determining the genetic diversity of wasps. This meant that physical samples had to be sent to the testing laboratories. It was decided that the participants could send in caught wasps for analysis, although in fact many of these were poorly preserved. By extracting DNA from the samples and using the polymerase chain reaction (PCR) technique, a number of small sections sited within genes could be amplified so that they could be analysed in detail for changes and variation between individuals and colonies. This technique of DNA extraction, amplification and analysis is straightforward when carried out in a fully equipped laboratory, so although citizen scientists provided the samples, the assistance of an academic laboratory was needed to complete the process. For the reasons mentioned earlier, wasps had generally been ignored by comparison with their close relatives the bees, and indeed ants, but the questions being addressed were just as important. When the analysis was complete the results yielded two important findings. First, the queens of *Vespula vulgaris* disperse across considerable distances from their original nest, thereby guaranteeing gene flow across the geographical range of the species and providing a uniform genetic population. Secondly, the distribution of trapped specimens indicated that foraging tended to be local to the nest. By the very nature of the study, these findings could only have been obtained by molecular analysis of individuals and by the well-organised collaboration of citizen and academic scientists. Although the final report did not mention how many citizen scientists had been stung by their quarry, it would be reasonable to assume that many of them were.

Although, as we have seen in the bee surveys, in some cases errors are easy to quantify and eradicate even if some data have to be abandoned, it is not always so easy. A good example of this can be seen in the analysis of errors associated with CoralWatch. This project aimed to define the extent and development of coral bleaching on coral reefs. Analysis of some of the older data threw up some interesting observations, and demonstrated the difference in focus that can be found between external users of the data and the people who collect it. In CoralWatch, the divers are involved because they care about the coral reefs, and may consider other peripheral data to be uninteresting and unimportant. It was found that the biggest category of errors was in provision of GPS data, but there were many other error categories, too, such as incorrect names and email addresses of volunteers. Other errors are probably more important, but may be a reflection of defective equipment, such as digital thermometers giving an erroneous temperature.

This clearly shows that there are many aspects of data collection where errors can be introduced. Interestingly, although these are often quoted for citizen science, they also apply to many other scientific endeavours. They can be broadly categorised as data collection not being clearly defined, the procedure not being followed correctly or the data being unsuitable for the investigation. As citizen science has gained traction and the various pitfalls of data collection have become apparent, so project design has become more sophisticated in an attempt to minimise errors.

Since the end of the nineteenth century, when science became a professional activity, it has become ever more dependent on technology. Sometimes this is to reduce workloads, but it is mainly to help to make measurements both easier and more accurate. Professional science has traditionally been funded from the public purse, but in the late twentieth and twenty-first century, funding has started to come from charitable groups and big business. It is important to view citizen science as a means of putting pressure on legislators. Rather than just telling people that the number of butterflies is decreasing, it is much more effective to get individuals to count them for themselves. This way conclusions will be drawn that can be presented so that they influence policymakers. Such projects can also last many years, if not decades, reflecting a time scale that professional projects rarely attain. Long periods of investigation are ideal for projects that are investigating biological processes, where change can be slow and may occur in very small increments. It has also been suggested that large citizen science projects with large numbers of volunteers are one of the likely ways for rare or anomalous observations to be made and verified – these may be important in the long term, and have more far-reaching effects than the original small project.

Citizen science is highly relevant to our understanding of the natural world, but more than that, it helps to improve the public understanding of science. It can bring a better understanding of the complexities and difficulties involved in aspects of science where it is assumed by many that there is always a clear and sound answer to any question that can be asked. Of course, science is not like that and it has its own ambiguities, so any data will add to the understanding of a subject, and may help to clarify an answer that is often fluid or uncertain. This is sometimes difficult to explain in an era that assumes all questions have a definitive answer, like a quiz, whereas in reality this is not the case. It is here that the true power of citizen science lies – in helping to understand a problem fully, in all its biological unpredictability.

Chapter 4

Small Science

What is small science?

We have looked at the long history of individual amateur scientists, well known because their ideas and studies led to big advances in established fields or even created new disciplines. During this period there were many more amateurs who observed and experimented with life science subjects in small ways as individuals or in local groups. Their work is mostly untrumpeted, lost in the huge body of scientific

FIGURE 4.1 Citizen scientists in action: surveying a bed of sea grass. (Philip Price, Seawilding)

knowledge or known only to specialists in the field to which it relates. Nevertheless, it has been, and can continue to be, a significant contributor to the giant on whose shoulders the professional scientist stands. In fact amateur science has its tentacles almost anywhere you care to look – in every branch of the life sciences, across all habitats and taxa, contributing to all levels of expertise, often in collaboration with the professionals and sometimes even taking the lead. Amateur scientists contribute to the phase of science that Kuhn described as 'normal' – those periods between the peaks of 'revolutionary' science, when slighter puzzles or problems are solved as scientists explore their chosen discipline within the confines of accepted theories. 'Small science' is the term we use to describe the puzzles or problems that are investigated by individual citizen scientists or local groups. It is more concerned with ecology, life history, behaviour and taxonomy, whereas 'big science' tends to focus on distribution and abundance.

There are several reasons for the obscurity of small science results. Sometimes they contribute only a tiny facet to an esoteric subject, or simply confirm that an already accepted hypothesis fits a slightly different set of conditions. Often they are insufficiently replicated to satisfy scientific rigour. There is nothing wrong with this if they are providing a small amount of supporting evidence or they prompt professionals to look more closely at a question. As James Lovelock says in his autobiography, *Homage to Gaia*, 'the first experiment in a new subject should be a rough one and intended only to get a feel for it' (Lovelock 2000: 129). A major hurdle for small science is the circulation of its results. Publication has always been a problem for amateur scientists. In the early days, before the mid-nineteenth century, results were shared by letter, and for a wider audience the scientist needed to be well connected or well known, with the time and money to write and send numerous letters, and in turn to receive and respond to replies. With the formation of clubs and societies it became possible to present papers at formal meetings, and these were then summarised and added to the proceedings published by the societies. Moreover, the societies would often encourage experimentation by their members, whether amateur or professional. For example, the Council of the Botanical Society of London (BSL) invited communication 'from members and others relative to Physiological and Systematic Botany' to be published in their proceedings. Daniel Cooper, the 19-year-old medical student who convened the meeting that established the BSL, was one of the first to communicate such a paper, which was published in the society's first proceedings in 1839.

Cooper's paper reported on two aspects of plant physiology – the effect of light on plant growth, and the uptake of water by roots. We know that his experiment was carefully designed, as he describes his methods in detail. He placed broad bean seeds in pots containing soil or an inert sand, grew them in the light or in darkness, and watered them with either plain water or water dyed red with madder, and then carefully observed what happened. He had hypothesised that because the beans that grew in the dark would be 'blanched', if they took up the dye they would become coloured or, as he put it, they 'might exhibit a singular appearance'. He had chosen madder because as a vegetable dye he believed it would not be toxic to the plants. He found no evidence of uptake of the dye even under the microscope, but made no further comment about this observation. More interesting is his very detailed description of the etiolated plants grown in darkness, compared with the de-etiolation caused by exposure to light. Plants grown in the light were green, with short internodes and fully expanded leaves, whereas those grown in the dark had no green pigmentation, and had long internodes (longer than those of light-grown plants by a ratio of three to one) in which the cells were much larger, and unexpanded and rudimentary leaves. Having satisfied himself that his observations were complete, Cooper switched the plants around, placing those that had germinated in the dark in the light, and vice versa. Both sets of plants withered after a few days, which in the case of the plants that had germinated in the light was not surprising. The dark-germinated plants had begun to show signs of de-etiolation, indicated by the greening of their rudimentary leaves, but they had probably exhausted all of their resources by the time Cooper switched them round. One hundred years or so later scientists used the de-etiolation response of beans and other plant species as models in their search for and eventual discovery of phytochrome – the pigment in plants which detects light and triggers de-etiolation and other photomorphogenic responses.

There are also many citizen scientists who, with hindsight at least, made fools of themselves by over-reaching their conclusions. For example, in the same issue of the *Proceedings of the Botanical Society of London* that included Cooper's paper, on pages 54–55 a Mr Woods wrote about the influence of barometric pressure on plants, noting the correlation between the stature of plants, barometric pressure and altitude. He observed that 'the effects of barometric pressure on the ... plant kingdom ... are plainly discovered in the instances of ... plants found on mountains and other elevated situations'. In science in general, and in ecology in particular, conclusions based on correlations alone are on shaky ground. We must assume that Mr Woods was unaware that the tree line decreases with increasing latitude whereas air pressure does not, because that knowledge would have rather blunted his hypothesis. In fairness to him, he also concluded that botanists would benefit by combining research in botany with that in meteorology, which of course is now a major part of ecological research. Also today, space scientists who have grown plants in space have found that air pressure does have an effect on plant growth. Other things being equal, lettuce plants grown at an air pressure equivalent to that at 5,000 m altitude are shorter than those grown at sea-level air pressure, although their weight at harvest is the same. However, this effect of air pressure is dwarfed by the effects of the changing seasonal temperatures and other factors such as snow cover and wind speed that increase with altitude.

Individual small science

Towards the end of the nineteenth century, in April 1897, a paper describing the observations of another citizen scientist was read to the Linnean Society, the same society to whom the papers of Wallace and Darwin concerning natural selection were first presented. This paper described the microscopic study of the germination and growth of spores of the fungus *Collybia* (now *Flammulina*) *velutipes*. As was the

FIGURE 4.2 One of Beatrix Potter's fine mycological illustrations, in this case a detailed watercolour of the Scarlet Waxcap *Hygrocybe coccinea*. (Armitt Museum and Library)

practice at the time, the paper was not read by the author. In this case it was read by George Massee, then Keeper of Lower Cryptogams at the Royal Botanic Gardens, Kew, and a Fellow of the Linnean Society. It was reportedly well received, with only a few modifications requested by the Fellows of the Society before it could be published. However, it was never resubmitted and no copies now exist. Nonetheless, we can be in no doubt that the paper was accompanied by exquisite, faithful illustrations drawn by the author, because the author was Beatrix Potter.

Before she became famous for her children's books, Potter had become a very capable amateur mycologist, an accomplished microscopist and a gifted artist. Her family were well off, and she was privately educated and encouraged to develop her interest in art and nature by her parents. By the age of eight she had started to keep journals with illustrations of the natural world she encountered, often on family holidays in Scotland. Her first sketchbook has passable watercolours of various moth caterpillars, including tiger moth and drinker moth. It was while on holiday in Dunkeld that she got to know a local postman, Charles McIntosh, who was also a citizen scientist – a self-taught naturalist who specialised in fungi. In fact she knew McIntosh from the age of four and it was probably he who ignited her interest in mycology. Later she corresponded with him on the subject and he collected material for her to work on. She experimented with spore germination in her kitchen, observing through her microscope and recording what she saw with very detailed scientific illustrations. She was the first to see and illustrate some aspects of spore germination, and she had grown spores of around 40–50 basidiomycetes species before she selected those of *Conocybe velutipes* for her paper. By this time she was in her thirties and had become known and respected by several members of the mycological establishment, including George Massee, who she had been consulting since being introduced to him by her uncle, who was vice-chancellor of the University of London. As a woman she was not allowed to become a fellow of the Linnean Society, and so was unable to attend to hear her paper read. It is clear from her journals, which have now been decoded, that she experienced misogyny in her interactions with the male scientists at Kew and the Linnean Society – she wrote that 'it is odious to a shy person to be snubbed as conceited and under the temptation of sauciness' (Potter 1989). Soon after withdrawing her paper, Potter gave up all her work on fungi, even her contact with McIntosh, and focused instead on children's books. No doubt the misogyny influenced her decision. In 1997 the Linnean Society publicly acknowledged that she had been 'treated scurvily'.

Birds have always been an attractive subject for lone citizen scientists, particularly the social behaviour of colonial species that live alongside humans, perhaps because they are relatively easy to identify and can be readily observed wherever one lives, even in cities. The popularity of ornithology means that it has been well supplied with serious publications, so it is easier for amateur ornithologists to have their work published. Unfortunately, this did not help James Burkitt, who published a series of articles in *British Birds* between 1924 and 1926 in which he reported observations on his local population of European Robin *Erithacus rubecula*. These were not widely read by the ornithological establishment. Burkitt lived in County Fermanagh, Northern Ireland, for most of his adult life and was by trade an engineer, being County Surveyor for 42 years. When he took up ornithology at the age of 37 in 1907, his approach was original, perhaps because at that time he did not know any other ornithologists. He was the first person to use coloured bands (in fact just black and white) on the robins' legs so that he could identify individuals and study their behaviour. He made the bands himself and adapted old wire-gauze meat covers to trap the birds prior to ringing them. The traps were sprung manually, with Burkitt waiting patiently for the robin to enter the trap, watching it out of the corner of his eye, before pulling away a supporting stick by means of a thread that was 30–100 yards long. He might have to stand for an hour or more waiting for a robin to enter the trap. That is dedication!

Over four years of study, Burkitt ringed and followed the fate of 88 birds, gaining some fascinating insights into the ecology of the robin. Most of his discoveries were new, although some had been guessed at. For instance, it is not just the male but also the female that sings, the pair working together to disguise the location of the nest. Burkitt wrote that 'Its natural movements are thoroughly deceptive'. The male may flaunt himself while the female skulks low along drains and undergrowth, or the pair may perform criss-cross movements. Pairs hold joint territories during the breeding season, and individual territories of about half the size thereafter. Watching individual robins allowed Burkitt to describe their moult in great detail. It starts on the chin and chest and progresses to the wings and tail. This part of the moult takes two to three weeks. There is then a resting period that lasts for a similar period before the head moult takes place. After the moult, when the robins make themselves known again, there is a preponderance of males holding the territories, and Burkitt hypothesised that it is mainly the females that migrate out of the country over winter. Finally, his observations allowed him to estimate the average lifespan of the robin as between one and three years. This is of course an average, and in a later study he recaptured a female 11 years after first ringing her.

At the end of the first paper in his series of robin studies, Burkitt encouraged other ornithologists to engage in this rewarding if time-consuming method of study, pointing out that he found the time for it even though he was not a man of leisure. However, he was rather ahead of his time, and his studies mostly fell into obscurity in the British ornithological community. However, he did inspire two citizen scientists who became very influential, one of them even turning professional. In Britain, David Lack picked up Burkitt's work on the European Robin when he was a schoolteacher at Dartington Hall School, where he taught from 1933 to 1939. He had been a keen birder since childhood. Indeed, by the age of 18 he had already won two national prizes for essays he had written about birds. He took a biology degree at Cambridge, was a member of the Cambridge Ornithological Society and published his first scientific paper on the display of nightjars when he was 22, so clearly his preferred career path was going to be ornithology. However, his first job was at Dartington Hall, where he was fortunate enough to have the encouragement and freedom to spend time studying the local robins by following individual colour-ringed birds. Like Burkitt, he studied their territorial and breeding behaviour and their song. Over a four-year period of field work he gained enough understanding to write the classic study *The Life of the Robin* (1943), a book mainly for the layperson, which has been enormously successful, with several editions being published, the most recent in 2015. Lack also wrote a two-part scientific paper, published in 1939 in the *Proceedings of the Zoological Society of London*, based on the same fieldwork. *The Life of the Robin* was published in 1943, while Lack was serving in the military during the Second World War. He never returned to the school, and in 1945 he turned professional, becoming Director of the Edward Grey Institute of Field Ornithology (EGI) at the University of Oxford. Lack's study of robins is also interesting because it is an early example of small group citizen science, where other interested people help with data collection. In Lack's case, the citizen science assistants were his staff colleagues and his pupils. He started the project to entertain and engage his pupils, but before long many staff and pupils were actively helping to trap and ring the birds and conduct surveys.

Somewhat earlier, a classic scientific paper was published in the *Transactions of the Linnaean Society of New York*. It was so large, at 620 pages, that it was published in book form in two separate volumes – the first, in 1939, on population studies and the second, in 1943, on the behaviour of the (North American) Song Sparrow *Melospiza melodia*. Nobel Prize-winning behavioural scientist Konrad Lorenz believed it to be the first such study of individuals of any free-living wild animal. The author was citizen scientist Margaret Morse Nice. She was born in Massachusetts in 1883 and was another youngster who was encouraged into natural history by her parents and became totally besotted with it. By the age of 13 she was making notes on

A. BEHAVIOR OF THE SONG SPARROW WHEN ALARMED

Three stages of fear may be distinguished in this species:

Stage	Note	Postures
"Alarm"	*tchunk*	Crest raised; tail raised and flipped; wings flipped; restless change of location.
"Fear"	*tik*	Feathers compressed; neck elongated; body crouched.
"Fright"	*tik-tik-tik*	Flies and hides; if caged, flutters in attempt to escape; pants with open bill.

FIGURE 5. *Attitudes of Song Sparrows in Alarm, Fear and Fright. From left to right (upper row): 1, unalarmed; 2, alert; 3, turning head to look; 4, 5, alarm (wings and tail flipped); 6, strong alarm; 7, fear; 8, fright (panting).*

FIGURE 4.3 An extract from Margaret Morse Nice's huge paper published in 1943 in the *Transactions of the Linnaean Society of New York*, in which she described her studies of the life history of the North American Song Sparrow. This diagram, drawn by Morse Nice, shows her interpretation of the various stages of alarm of the sparrow when responding to the presence of an avian predator. (HathiTrust)

the behaviour of her sister's hens, regarding 'who pecked whom'. She did start to pursue an academic scientific career, studying child psychology and zoology at university, but gave up her professional studies after marrying in 1909. Looking after a growing family – she had five daughters – and moving around the country in pursuit of her husband's career was not conducive to her own career. However, these family commitments did nothing to dull her enthusiasm for scientific study, and she spent as much of her spare time as she could (and indeed some of her family time) investigating the subjects that were closest to her heart. During her busiest years as a mother her studies focused on two of her daughters, whose developing vocabulary she recorded in great detail, and she later published 18 articles on speech development in children, combining her knowledge of child psychology with her daughters' data.

Once the children were older and she had more time, Nice returned to ornithology as her main field of interest, and when the family moved to Columbus, Ohio, in 1927 she started what was to become her seminal study on the Song Sparrow *Melospiza melodia*. Like the House Sparrow *Passer domesticus*, the native Song Sparrow is common, widespread, relatively sedentary and tame, making it a very convenient subject for study. Nice observed the song sparrows on the family's 24 ha (60 acre) property, where 'there is a large population right at my door'. She was influenced by the work of Burkitt, and like him used coloured bands to enable her to follow individual birds. Like Burkitt's, much of her kit was homemade. She used a mixture of homemade and 'government sparrow traps', and, until 1932, when they became commercially available, made her own bands from brightly coloured celluloid poultry bands. She cut these down into suitably sized strips which she wrapped around a nail, held in place with an aluminium band and heat-cured in boiling water. During the eight years for which she studied the song sparrow she ringed more than 800 of these birds, but she spent most of her time observing those that held territories close to her home. For example, on 11 May 1935 she followed one male sparrow for 24 hours, during which period he was sleeping for nine hours, singing for ten hours and eating or otherwise engaged for five hours. The upshot was a comprehensive study of the song sparrow's life history, covering song, nesting, territoriality, survival, lifespan and social behaviour. The work was actually first published in German some years earlier than the English version, in the *Journal für Ornithologie*, because Nice had had trouble getting such a long article published in an American journal. The song sparrow work brought her international recognition in the ornithological world, but it was by no means her only contribution to ornithology. She studied the local bird species wherever she lived, and during her lifetime she published 250 articles on birds in scientific journals. She was awarded the American Ornithological Union's Brewster Award, made an honorary member of the ornithological societies of several countries and received two honorary doctorates.

The House Sparrow is believed to be the commonest bird in the world, despite recent population crashes in western European cities, and it is the archetypal avian companion of the human race. Common, tame, widespread and colonial, it is the ideal subject for amateur study, or at least Denis Summers-Smith (1920–2020) thought so when he watched house sparrows flying in and out of his hospital window as he lay recovering from war wounds in 1946. He was a mechanical engineer by profession, an expert in tribology (the study of lubrication, friction and wear of mechanical moving parts), but in the ornithological world he is known as the sparrow man. He became one of the foremost experts on the 27 or so species of Old World sparrow in the genus *Passer*, and the house sparrow in particular. His job took him all over the world, and wherever he went there were always sparrows close by for him to observe. Such was his passion that he had a few brushes with the law: he was arrested twice for watching sparrows in people's gardens in the early morning, and he was moved on by the authorities when sparrow watching near a rocket launch site in Turkestan. By the end of his life he had written five books, one of which is still the definitive monograph on house sparrows, and numerous scientific papers. He had also helped to establish and then served on the Granivorous Birds Committee for the International Biological Programme, and had received the Zoological Society of London's Stamford Raffles Award.

Summers-Smith's early studies were undertaken around his home village in Hampshire, as post-war petrol rationing severely limited his ability to roam. He identified and observed 25 house sparrow colonies in his village, and colour-ringed the birds of one colony so that he could study them in detail. He recorded and drew vector maps of flight direction for all birds he saw flying for more than 20 yards (18 m). He was also able to ascertain the timing of certain events that he could not see directly. For example, he could not determine the date of first laying by direct observation because most nests in the colony were in cavities in walls or under roof slates or in roof thatch, but he could tell by behaviour from his experience of watching the sparrows nesting in nest boxes that he had erected. He was careful to stress that his results applied to a rural setting, and to contrast his data with the few data from urban sites that were available at the time. The results are fascinating. Each colony had a core and a peripheral area. All nest sites in the core were always occupied every year, mostly by established pairs, whereas those in the periphery were sometimes left empty and had a tendency to be occupied by first-year birds. Once a pair had nested in a colony they were likely to remain faithful to that colony for life, so generally only young birds moved away, and then usually to other colonies. In fact, males would mostly remain faithful not only to the colony but also to a nest site, especially if it was in the colony core, and would visit the site every day throughout the year. All of this behaviour would tend to maintain the existing colonies and keep them 'tight'. Various behaviours would help to synchronise colony activities. For instance, Summers-Smith concluded that the communal display, popularly known as a sparrow's wedding, which involves the raucous chasing of a lone female by (eventually) as many as ten or more males is one of the signals that brings the whole colony into breeding condition. David Lack, who was by this time Director of the EGI and a Fellow of the Royal Society, encouraged Summers-Smith to publish his data, which he did in several papers in *British Birds* and *Ibis* in 1954 and 1956.

Clearly, relatively simple, detailed observations can yield some intriguing insights. Of course they cannot tell us everything, and they can be fallible, as one cannot study every animal round the clock every day, and Summers-Smith gives an example of this. During his many years of observing ringed house sparrows – often, as he put it, for 10% of the sparrow's waking hours – he never once witnessed a female being unfaithful to her mate, and he believed that they were monogamous. Yet when DNA profiling became available it was discovered that 10–20% of youngsters had a different father to the one that was rearing them.

Ringing continues to be a dynamic subject for small citizen science, as well as for big international projects, with a wealth of questions within the scope of small ringing groups still to be answered. A case in point, through which a thread runs from individual study to local group projects and on to national and international schemes, is the story of bird study in Nottinghamshire Wildlife Trust's Treswell Wood, which has been running for 50 years and is continuing. The driving force behind the study was John McMeeking (1929–2019), a local lad who became a businessman by profession, running a Nottingham lace-making company, but was at heart an ornithologist. He began ringing as a teenager with the help of his school biology teacher, and he still held his ringing permit when he died 75 years later. He was also a passionate conservationist, and in 1968 he became a founder member of Nottinghamshire Wildlife Trust. Four years later he assisted with the purchase negotiations for the trust's first large woodland reserve, Treswell Wood, which covered an area of 48 ha, and he set to work immediately ringing, surveying and, in his words, 'recording what was there'. This rather flippant remark belies McMeeking's plans for a detailed and integrated ecological study of the woodland's birds, and over the next few years he established a number of programmes and started to gather a dedicated group of volunteers to implement them. The first programme, initiated in 1973, aimed to make Treswell one of the woodland sites for the BTO's Common Birds Census, in which volunteer surveyors make eight or so visits to the same site every year during the breeding season to record the location, activity and species of every bird that they see or hear. Subsequently, experts at the BTO analyse the data and convert them into territory maps. In 1978, McMeeking started another programme which was in effect one of the three pilots for another BTO monitoring scheme, known as the Constant Effort Sites (CES) scheme. In this scheme, volunteer ringers set the same nets in the same places at the same times at a number of intervals during the breeding season every year, and they ring newbies, record recaptures and measure the birds' vital statistics. This work not only led to the first paper based on the Treswell studies, in 1991, but also helped to verify the protocols for the CES that was formally launched in 1983 and has since been emulated in Europe and the USA. Finally, a nest box monitoring scheme was started in 1979, with around 200 boxes being erected. These are monitored weekly during the breeding season. McMeeking was also very aware that the activity of birds would be shaped by their environment, and he made sure that any factors that might be relevant, such as the weather and habitat management activity, were also recorded.

Although McMeeking and his volunteers were well aware of the importance of their data and the questions that their findings might answer, they did not have the expertise to perform detailed scientific analysis of the data, so they collaborated with university academics for this purpose. This led to the publication of a dozen papers specifically on Treswell, with the data also being used in international studies. The Treswell studies have also been a useful resource for students, many of whom have been introduced to ringing. These studies nicely illustrate the value of small citizen science projects. It is extremely unlikely that a professional academic study could have provided the human resources needed over the 50-year time span of the project, and they contributed something which is impossible for big citizen science projects to achieve, namely a detailed investigation by a consistent group of observers. For example, an understanding of the winter survival of the Wren *Troglodytes troglodytes* and Treecreeper *Certhia familiaris* was gained by analysing 20 years' worth of systematic mist-netting records together with weather records for the same period. Survival in both birds depended on the severity of the cold, but in treecreepers it also depended on rainfall. A cold, wet winter is not good for treecreepers. For wrens, the results showed density dependency underlying the severe mortality caused by harsh winters. More wrens are lost when their numbers are larger to start with, suggesting either a higher mortality or permanent emigration. Nature never stands still, and population fluctuations have always occurred, often on a cyclical basis, but in these days of global warming we need studies like this to monitor more permanent changes.

The world of insects is also a rich source of projects for small science, and in many ways is also well suited to it. As we have seen, the broad brush of big science is excellent for looking at the distribution and abundance of the larger, more colourful, more easily identified insects, but when it comes to the more critical, hard-to-identify species, individual citizen scientists rule the day. In total they can afford to dedicate many more hours to the subject than the professionals, and for some groups it is the amateurs who are the experts, there being far more insect taxa than there are professionals to claim ownership of them. Many people would assume that we know all of the insects and other animals that live in Europe, and that the European biodiversity catalogue is complete, yet since the 1950s an average of 770 new species of animal have been described every year. This roughly equates to a 0.5% increase in the number of animal species in Europe every year, and almost 75% of these are insects. Importantly, amateurs have been responsible for describing a large proportion of these new insect species – approximately 85% of the grasshoppers, locusts and crickets (Orthoptera), 60% of the butterflies and moths (Lepidoptera), 55% of the beetles (Coleoptera), 50% of the bees, ants and wasps (Hymenoptera), 50% of the flies (Diptera) and 25% of the true bugs (Hemiptera). The situation is similar in the rest of the world. Without the amateurs, insect taxonomy and indeed biodiversity as a whole would be in a much poorer state, as would conservation. There is so much that we do not know about the distribution and life history of the new or critical species, and both of these subjects fall directly into the purview of expert amateur naturalists. If conservationists are only aware of the larger, more obvious fauna on their patch, such as the rare butterflies, then management for their benefit may compromise the health of the less prominent fauna.

Like ornithology, entomology is considered to be a proud tradition among citizen scientists. Perhaps one of the first of the modern school of entomologists was William Tutt (1858–1911). In the nineteenth century and earlier, most amateur entomologists were focused on collecting specimens and were especially keen to seek out the aberrant. The magnitude of their collecting, like that in other areas of natural history, such as bird's eggs, ferns and orchids, often caused significant damage. Tutt, a schoolteacher by profession and an expert amateur lepidopterist, was passionate in his objection to this approach, and advocated instead the study of the insect's natural history and trying to understand how the aberrations arose. He was one of the first to propose that the rapid increase in the dark (melanic) form of the Peppered Moth *Biston betularia* in industrial areas, where the trees on which the moth rests during the day had lost their moss and lichen cover and in some areas had even become covered with soot, was due to preferential predation of the more visible, normal, pale-coloured form by birds. This became a famous example of natural selection, though not evolution, taught in biology classes in schools and universities throughout the world. However, credit is usually given to a professional scientist, Bernard Kettlewell, who around 60 years later backed up the observations with experiments demonstrating that it is harder for birds to find dark forms of moths against sooty backgrounds.

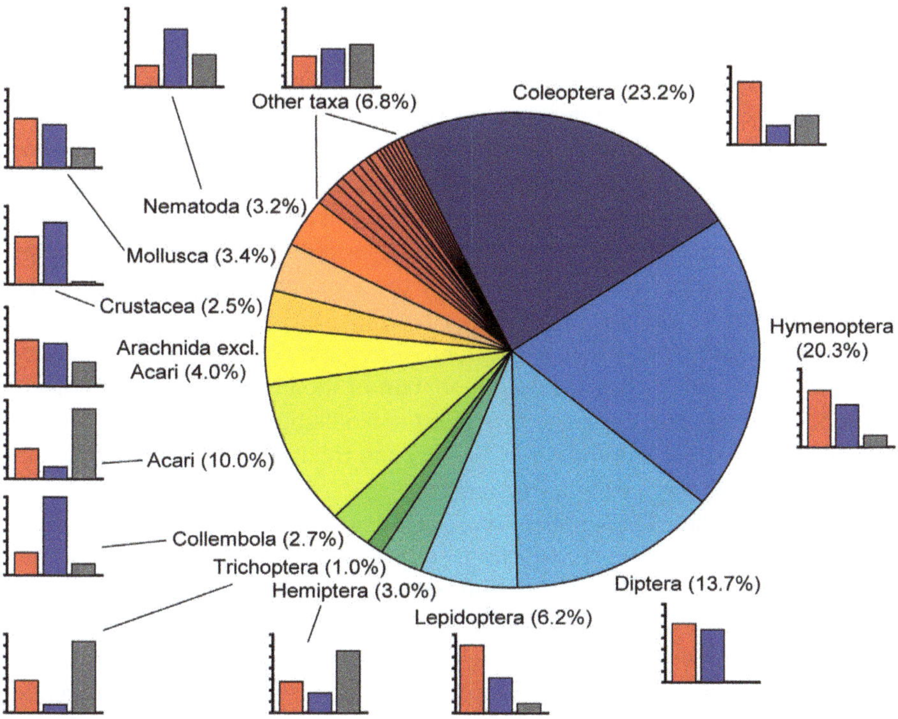

FIGURE 4.4 A pie chart showing the distribution among orders of the 5,881 new species of animal described in Europe between 1998 and 2007. The significance of the contribution of citizen scientists is clearly shown by the histograms, in which the red bars denote the proportion of species in each order described by amateurs, with the proportion described by professionals shown in blue, and unknown sources shown in grey. (Source: Fontaine *et al.*, 2012, available at https://doi.org/10.1371/journal.pone.0036881)

Cynthia Longfield (1896–1991), another notable twentieth-century entomologist, and probably one of the last moneyed amateurs, was keen on dragonflies. In fact she came to be known as Madame Dragonfly because of her expertise on the subject. She was educated at home by governesses and was a self-taught naturalist, collecting caterpillars to rear when she was on the family estate in Ireland and, when in London, taking specimens to the Natural History Museum for comparison with those in its collection. Her reputation grew and she was accepted as a self-funded amateur entomologist on an expedition to the Pacific Islands in 1924. This is where she first focused her interest on Odonata (damselflies and dragonflies). During the expedition she was helping to collect specimens for the Natural History Museum, and on her return she worked in that museum as an unpaid associate cataloguing the Odonata specimens. She continued to work there as a volunteer Odonata specialist until she retired at the age of 60. During this time she took herself on many more collecting expeditions around the world, became an international authority on the Odonata, published many papers, and wrote what would become the standard handbook, *The Dragonflies of the British Isles* (Longfield 1937). In her retirement she wrote *Dragonflies* (Corbet, Longfield and Moore 1960) as part of the Collins New Naturalist series, was instrumental in setting up the British Dragonfly Society, and could be seen with net, binoculars and walking stick on field trips around Ireland. It seems that Madame Dragonfly was a very apt moniker!

In the UK, expert amateurs are the backbone of field entomology. Facebook groups, through which anyone can post images of insects for identification, are screened by experts who are usually amateurs. It is often amateurs who set the standards and protocols and run the national recording schemes. For example, almost all of the 30 schemes for flies run by the Dipterists Forum are organised by amateur experts. At least half of the articles in technical journals and bulletins, such as the *British Journal of Entomology and Natural History* and *Dipterists Digest*, are written by amateurs. Amateur data are the foundation of the insect Red List. The deeper one looks the more one finds, and this is further highlighted by the close association between the Amateur Entomologists' Society and the Royal Entomological Society, which itself has a large amateur membership.

Consider sawflies. The adults of most of these flies are small and insignificant, rarely noticed by anyone except the keen entomologist with a sweep net. Gardeners and allotment holders will be very familiar with their larvae. We personally have witnessed a gooseberry bush being stripped of its leaves in a couple of days by caterpillars of the Common Gooseberry Sawfly *Nematus ribesii.* There are about 550 sawfly species in the UK, but less than a dozen county recorders or verifiers for them. The majority of these experts are amateurs. John Grearson is a county sawfly recorder for Wiltshire. Until he retired he was a management accountant. His entomological knowledge came from books, articles and fieldwork, and his interest in sawflies probably developed from his involvement in local nature conservation, when he would often be called upon to survey the insects in a particular area. His specialist interest is in the genus *Pristiphora*, to the extent that he has published notes on the identification and distribution of a new species for Britain, Lime Sawfly *Pristiphora leucopus*, together with a revision to the standard identification key to include it. For this he employed the entomologists' trick of collecting larvae and rearing them until they metamorphosed into adults, a method known as rearing out, so that the larval stages could be identified. Matching larva to adult can be quite problematic in entomology, and there are many examples where the adult insect is known and its larva is not. Grearson also added to the identification resources available to entomologists by posting his sawfly photographs on Flickr. This is now common practice among entomologists.

Steven Falk, probably a more well-known amateur entomologist through his *Field Guide to the Bees of Great Britain and Ireland*, maintains a vast Flickr account – an absolute gold mine for entomologists. At the time of writing he estimates that it contains about 70,000 images of 2,500 species across the nine orders he knows best, from beetles and earwigs to dragonflies and grasshoppers. He has taken almost all of the photos himself, and they include his own specimens, other entomologists' specimens and the collections of the British Museum and Oxford University Museum. Each family and species has a short description with pointers to other identification resources, and species images are accompanied by photographs of their typical habitat. Falk, who is also an accomplished artist, was drawn into the world of insects as a young boy through his art. At an Amateur Entomologists Society exhibition he was encouraged by a friend to 'go and see who that geyser is who keeps looking at your paintings'. The geyser was Alan Stubbs, who happened to be starting to write his book, *British Hoverflies* (Stubbs and Falk 1983). With the confidence of youth Falk asked if he could illustrate it for Stubbs, which is why, at the age of 15, he came to be a co-author and illustrator. He went on to do a zoology degree, then to work as an entomologist for the Nature Conservancy Council and in the natural history section of Coventry Museum, and he is now an independent ecological consultant. His holidays often include entomological studies. He spent six years of annual holidays, sweep net in hand, surveying the bees on over 100 chalk grassland sites on the South Downs in Sussex, noting not only the species but also the food plants and habitats they used. This nicely demonstrates how a detailed understanding of the ecology of a habitat can be gained by careful observation and determination.

Our final example of a contemporary amateur entomologist is David Chelmick, who until he retired was a quantity surveyor. He became interested in the Odonata (dragonflies and damselflies) by accident. As a young man in the late 1960s he had taken up ornithology as a hobby, volunteered to be a reserve warden for a week and ended up at Arne Nature Reserve in Dorset because the bird mecca of Minsmere in Suffolk had all the wardens it needed. It was the professional warden at Arne who introduced Chelmick to dragonflies, and he immediately became hooked. There were many challenges at that time – only scientific names were used, keys mostly included only drawings of part of the insect, such as wing venation or anal appendages, and unfriendly landowners and fishermen abounded and would occasionally give him 'a dip in the drink'. Nevertheless, he eventually became sufficiently expert to be made organiser of the Odonata Mapping Scheme in 1976, and he edited the second edition of the Odonata section of the *Provisional Atlas of the Insects of the British Isles* three years later. His focus of interest then moved overseas, first to Europe and then, in the late 1990s, to the rainforest of Cameroon, as part of the Cameroon Dragonfly Project. South-west Cameroon is the wettest place in Africa, and one of the wettest in the world, but Chelmick appears to have relished the mud and water, which must be a good thing if one specialises in dragonfly larvae. When the project started very little was known about the adult dragonflies of Cameroon, and nothing about their larvae. Over the eight years of the project its small team, with the help of local people, captured hundreds of adults and larvae, and reared out many adults. For them, rearing out was often the only way to unequivocally link adult to larva, because they found that the adults of many species spend all of their time in the rainforest canopy, with only the females descending to lay their eggs. In fact the larvae of *Cornigomphus* species were the third most common dragonfly larvae that they found, yet they never saw an adult. A fascinating find involved close examination of the larva of the damselfly *Pentaphlebia stahli*. The larvae of this species live on rocks in fast-flowing mountain streams and have a highly unusual morphology – an adaptation designed to help them to cling to the rocks – found in just one other species, which lives in similar conditions in Venezuela. The theory was that the two species shared a common ancestor which was alive when Venezuela and Cameroon were adjacent areas in the same landmass, before South America and Africa split apart. However, subsequent DNA evidence suggests that that is not the case, so their similarity is probably a result of convergent evolution.

Much of the work on plant distributions is also supported by citizen scientists. The majority of the species records are collected by amateur botanists, some of whom submit tens of thousands of records every year. Most of the county recorders are expert amateurs, as are about two-thirds of the referees. County recorders are the initial gatekeepers who validate the records submitted by field botanists before entering them into the distribution database, and referees are the specialists to whom specimens that are difficult to identify are sent for verification. However, many amateur botanists are involved in much more than just square bashing – that is, recording all of the species found in the monads (1 km squares) into which the country is divided. They undertake their own studies, sometimes analysing subsets of the records stored in the distribution database, and sometimes taking a more experimental approach to plant ecology. John Box is one such amateur botanist. He is in fact an ecologist by vocation and avocation. One of his studies as an amateur was sparked by fascination with the link between the festive season and the triumvirate of evergreens – Holly *Ilex aquifolium*, Ivy *Hedera helix* and Mistletoe *Viscum album* – and, through that, the rather rare association between Mistletoe and oaks (*Quercus* species). Gathering together records from a multitude of sources – the BSBI's distribution database, the Association of Local Environmental Records Centres, appeals he made to BSBI members, published sources such as floras and a dozen or more herbaria – he found that there are just 13 mistletoe-oaks today, and that this low number has been fairly stable, varying between 10 and 20, over the nineteenth and twentieth centuries. The age range of mistletoe-oaks alive today, which he estimated from their diameter at breast height, is between 30 and 400 years, and it may be that the mistletoe, once established, lives for a similar age. Since the association is so rare, individual trees are mentioned in historical accounts, and for one oak there are records to suggest that its mistletoe has been growing on it for 150 years.

David Trudgill is another example of an amateur making use of the big data stored in the BSBI's plant distribution database. He became interested in native orchids after retiring from a career in zoology, and he wanted to use the distribution database to re-examine the changes in the occurrence of native orchids in Scotland, where he lives. Previous analyses were somewhat equivocal as to whether the distribution of some species had actually increased. There are two major problems with the distribution database. First, recorder effort has increased hugely over time. If one just looks at the raw data by counting the number of tetrads (2 km squares) in which an orchid species was found (orchid tetrads), the occurrence of all of the species appears to have increased over time. This problem can be largely overcome by using the number of orchid tetrads expressed as a percentage of the total number of tetrad records for all plant species. When this adjustment was made, 19 of the 20 species showed a decrease in occurrence, and two-thirds of these decreased by more than 50%. The second problem is that these are occurrence data. Orchid numbers may have declined dramatically in a tetrad, but so long as a single plant remains the tetrad will be counted. Nothing can be done about this, as the data simply do not hold that information, but it does mean that the decline could be even more severe than the analysis suggests. Trudgill performed one further analysis of the data, the results of which are rather intriguing and quite worrying. When all of the orchid species are considered together and the decline is extrapolated, it shows orchids completely disappearing from Scotland in 2052. This is very unlikely to happen, as such extrapolations involve so many assumptions and unknowns, but it shows that there is no room for complacency when combating biodiversity loss.

The problem of monitoring change in frequency of species rather than just occurrence is not as intractable as it might seem, and there are a couple of botanical citizen science projects designed to address this. One solution is to reduce the scope by not trying to cover the whole country. This is how the BSBI's National Plant Monitoring Scheme works. Volunteer botanists carry out an annual survey of the same few 5 × 5 m plots they have chosen within a monad to which they have been assigned. Another solution, which is probably only feasible if an area is sufficiently important to motivate the work involved, is to collect records on an extremely fine scale. This has been done on Cronkley Fell in Teesdale, a unique area in Britain where a large collection of rare alpine plants grow together – the upshot of history, geology, land management and a surprisingly modest altitude. The driving force behind the survey is Margaret Bradshaw, who has been championing the Teesdale flora for almost her entire life, both professionally and as an avocation. Even at 98 she is still doing charity walks to raise money for her Teesdale projects. For Cronkley Fell, the plan is to survey every single square metre, and so far a professional botanist and a team of volunteers have surveyed over 320,000 of them.

Trudgill and Box have also conducted more experimental studies, both of which, as it happens, are about native orchids. Many temperate orchids become dormant over winter and need chilling in order to break dormancy in the spring by a physiological process called vernalisation. Trudgill investigated the effect of winter chilling on marsh orchids (a mixture of Common Spotted-orchid *Dactylorhiza fuchsii* and Southern Marsh-orchid *Dactylorhiza praetermissa*) and Lesser Butterfly-orchid *Platanthera bifolia* by keeping pots containing adult tubers taken from path edges in his orchid meadow in an unheated outbuilding or indoors in a cool room where the temperature never fell below 9°C. Both sets of plants were moved to the same area outdoors in early spring. The experiment did indeed show the need for vernalisation in these species. Although the plants that were overwintered outdoors developed buds more slowly, they grew into robust plants, whereas those that were overwintered indoors grew more slowly and were stunted. Box's study used counts of the same two marsh orchids as a measure of the restoration of an urban-edge field into meadow, and he recruited 20 or so local people to collect these data. Every year between 2005 and 2021 they would quarter the field, walking in a line each holding a knot separated by 1.5 m on a length of string, and count the number of orchids in flower to their right as far as their neighbour on the string. The number of flowering spikes increased from 19 to 3,338 over the 17-year period. A plot of the number of spikes against time fits an S-shaped curve, suggesting that the maximum is likely to plateau at around 4,000. One can also conclude from this that, as with regeneration of woodland, meadow restoration takes decades.

Some of the most engaging small science projects are those that look at species interactions. The pollination of plants by insects falls into this category. Victor Chambers (1911–1984), a research chemist by trade, was an avid naturalist whose interest focused on insects, particularly sawflies. Such was his expertise that he has a sawfly named after him, *Rhogogaster chambersi*. However, his passion for insects started with bees, and he wanted to find out which plants were used as pollen sources by the species he was interested in. This involved collecting pollen from the most likely plants, finding the bees' nests, capturing the bees on their return to the nest and putting them in test tubes, waiting patiently for each captured bee to clean the pollen from its body and deposit it in the test tube, and later comparing the bees' pollen with his own pollen collection. He was apparently painstaking in ensuring that he released the bees at exactly the same spot where he had caught them.

Greater Butterfly-orchid Platanthera *chlorantha* is pollinated by night-flying moths. It has a white flower, the scent of which is strongest at night, and its nectar is produced in a long thin spur at the back of the flower. To sip on the nectar the moth must push its proboscis down the spur until it reaches the nectar. Orchid pollen grains are not released separately. They are moulded into two tiny blobs called pollinia, each of which has a stalk and a sticky disc or viscidium. When an insect pushes against it, the viscidium glues the stalk and pollinium to the insect, and so the pollen blob is transported to the next flower that the insect visits. Therefore, for pollination to be successful in Greater Butterfly-orchid, the moth's proboscis must be just the right length for it to have to push against the viscidia before it can sip on the orchid's nectar. If its proboscis is too long it will be able to drink the nectar without carrying off the pollinia, which is of no benefit to the orchid, and if it is too short it will not be able to reach the nectar, and although it may carry off the pollinia, it may be disinclined to try other flowers of this species, which is of no benefit to either the moth or the orchid. Furthermore, the morphology of the flower of the Greater Butterfly-orchid must be such that the viscidia make contact with the moth's eyes, because the flaky scales that cover the rest of the moth's body would prevent the viscidia from sticking firmly. Roy Sexton, a retired academic plant physiologist, and a group of volunteer moth trappers set out to see if this theory holds true for their local central Scottish populations of Greater Butterfly-orchid. They used a light trap to capture moths, looked for any with pollinia from this orchid species attached, and measured proboscis length. Sexton also measured the average length of the orchid's spur to the surface of the nectar. As expected, spur length and proboscis length were very similar. However, relatively few moths were found with pollinia attached to their eyes – the maximum number caught in an evening was 15. Given this low figure, the short flowering season (two to three weeks), the brief five hours of darkness at that time of year in central Scotland, the moths' reluctance to fly in poor weather and the huge number of individual flowers to be visited (5,000 or more in large populations of Greater Butterfly-orchid), Sexton has questioned how widespread moth pollination really is.

Small science in groups

The modern perception that group citizen science is a new trend is incorrect. In fact it goes back a long way, and not just because citizen scientists tend to be social beasts with a similar interest, but also because it has always been the case that some studies are not practicable for the sole researcher. The view that group citizen science is new probably stems from the fact that such studies now have the citizen science label and their existence and outcomes can be easily publicised on social media. Previously, a study supported by a band of interested volunteers could be relatively obscure. For instance, the help that David Lack received from other teachers and pupils in his study of robins is not obvious from his publications. Another example comes from Margaret Percival's study in the 1940s of pollen collection from a range of wildflowers by the Honey Bee *Apis mellifera*. Percival was a professional scientist working at the University College of South Wales and Monmouthshire, now Cardiff University. Her study involved making repeated counts throughout the day of the number of pollen loads from different wildflower species that the bees brought to their hive. The count was done visually. Unlike Chambers, who had to physically examine the pollen on the bees, Percival found that the species from which a bee had collected the pollen could be distinguished by the colour of the compacted pollen in its pollen sacks and the distribution of pollen on its body. At the same time a map was created showing the distribution and abundance of the major plant species within a half-mile radius of the hive and their peaks of flowering. In the paper's acknowledgements Percival thanks 'the Radyr Girl Guides for their assistance in the floral survey'.

Similarly, natural history societies have been bringing likeminded groups of naturalists together for many years. At the time of writing there are nearly 90 societies in the UK. The Ashmolean Natural History Society of Oxfordshire (ANHSO) is the oldest in England, with a history that can be traced back to 1828. It was developed into its current form by George Druce when he engineered the merger of the original Ashmolean Society with the Oxfordshire Natural History Society in 1901. This is the same controversial Druce who both re-invigorated the Botanical Exchange Club and was accused of over-zealous collecting. Like all natural history societies, the ANHSO's aim has always been to promote the study and enjoyment of local nature, and it has run many projects throughout its history. In 1907, for instance, it published a preliminary list of the beetles (Coleoptera) found close to Oxford. This list, which totalled 1,400 species within the 7-mile (11-km) radius of the search area, was put together by James Walker, a retired naval officer, with the help of several other enthusiastic entomologists, three of whom worked in the University Museum.

A more recent publication by the ANHSO describes another survey of beetles. This one took place over a five-year period, from 2014 to 2018, in an ancient beechwood in the Chiltern Hills. There is historical documentary evidence that Lambridge Wood has been in existence since before 1600, and natural evidence points to its antiquity, too, with the presence of Bluebell *Hyacinthoides non-scripta*, Wood Spurge *Euphorbia amygdaloides* and Woodruff *Galium odoratum*, all of which are indicators of ancient woodland. However, it is semi-natural. It was a working wood until the middle of the twentieth century, the beech trees being used originally to provide wood for the local furniture trade. When this industry declined they were used to make tent pegs in both world wars, and finally they were used for brush heads. Ancient semi-natural habitats can be very diverse, and indeed the surveyors found around 250 species of beetle, 20% of which are local or notable. All manner of collection techniques were used, from pitfall traps, sweeping and beating to an ethanol-baited trap that was suspended to catch beetles on the wing, and the use of a cherry picker for canopy sampling. Many different habitats were sampled – flowers (for pollen and nectar feeders), moss banks, rotten wood, flooded knot holes and deer droppings, among others. The survey can form a baseline for future studies on the effect of climate change and can also help to determine the status of other Chiltern woodlands whose histories are not so well known.

The project is of course compelling in its own right, but it is particularly interesting to us because it brought together a very varied group of citizen scientists. The lead in the study, Richard Fortey, who actually owns a small part of the wood, is a lifelong career palaeontologist with the Natural History Museum. He was joined at various points by Coleoptera specialists (also from the Natural History Museum), schoolchildren from Minibeast Mayhem and interested members of the general public on a bioblitz day. Minibeast Mayhem, which is a company that specialises in teaching entomology in schools, was set up by Sally-Ann Spence, an expert entomologist and an honorary associate of the Oxford Museum of Natural History. On the day of the bioblitz, Sally-Ann Spence brought along about ten enthusiasts. A bioblitz is a race to discover as many species as possible at a set location over a given time period, usually a day. It brings together scientists, students and members of the public. Its aim is to generate interest in biodiversity by injecting excitement and fun into the search – the idea of a race, keeping it local and offering the opportunity to engage with professional scientific experts. This worldwide citizen science initiative started in the USA in the late 1990s with a bioblitz held in Kenilworth Aquatic Gardens, Washington DC. The UK National Bioblitz Network is hosted by a charity, the Natural History Consortium, and as well as organising its own events it offers free resources to help others to do so. The Natural History Museum is just one of many organisations in the UK – including councils, wildlife trusts and universities – that hold bioblitzes.

Turning our attention to aquatic habitats, and moving towards the sea, beachcombing can tell us a lot about what is happening offshore. This is an area where the citizen scientist tends to reign supreme. Much time needs to be invested and finds are made by chance, so the professional scientist can use the records posted or submitted by the amateur and help the amateur with identification, but it would not be practicable for them to do all of the searching. Beachcombing is an activity suitable both for individuals, who can become real experts, and for local groups that welcome all comers. Husband-and-wife team Steve Trewhella and Julie Hatcher are good examples of the former. Although Julie is a professional marine biologist and Steve is a wildlife photographer, most of what they do 'is pure interest, or to highlight conservation issues', and they seem to spend most of their spare time beachcombing or diving. Their finds are fascinating and often miniscule. A toilet brush that they found washed up on Chesil Beach, when finely dissected, revealed two travellers among the goose barnacles that had also made their home on the brush, namely the sea slug *Fiona pinnata*, which has a worldwide distribution, and a specimen of the tropical Columbus Crab *Planes minutus* measuring 2 cm across. All three species are pelagic, living on floating objects in the open sea, which increasingly these days are plastic items. The Columbus Crab, which more typically is just 1 cm across, probably started its journey in the Sargasso Sea, perhaps on *Sargassum* weed or a turtle, its more normal, natural home. The crab swims out short distances to catch its food (small fish, krill and squid) and sometimes to move from one host plant or animal to another. At some point it must have selected the toilet brush and then drifted with the Gulf Stream to the Dorset coast. Columbus crabs are rare in northern Europe, but Steve and Julie have been finding them more frequently in recent years. When they found their first specimen, around 15 years ago, it was the first Columbus crab to have been seen in the UK for 200 years. The increase in recent years may reflect the warming waters and the increasing amount of plastic debris in the seas. Plastic objects have a much longer life than the organic debris that these tiny pelagic ecosystems would naturally inhabit, and they thus have the potential to carry epibionts (pelagic organisms that spend their lives on floating debris) much further, beyond their natural range, to areas where they could become invasive. Scientists at the Smithsonian Environmental Research Center in the USA recently gathered data to evaluate the extent of the plastic-assisted spread of epibionts – another example of a study that was only feasible through citizen science. Around the world more than 1,000 participants returned data from around 430 beaches.

Mermaid's purses, which are the egg cases of sharks and rays, get washed up on to the strandline and are much more common finds on beaches. They, too, are the subject of citizen science recording in the UK, which is often organised by local communities. A very nice example of this is the village of Penparcau just outside Aberystwyth in Wales. The local people have been recording counts of mermaid's purses on their beach in a systematic way since 2015, running 11 searches per winter season. Five species are commonly found – two sharks and three rays – with 20–100 egg cases of each of four of these species usually found over the course of a season. However, one of the shark species, the Nursehound *Scyliorhinus stellaris*, regularly has 500–1,000 egg cases washed up per season. The conservation status of this species has been assessed as Near Threatened, and the huge number of egg cases that the Penparcau citizen scientists find every year suggests that Cardigan Bay is one of its nursery grounds. Before the Penparcau surveys there were very few records of sharks and rays in the area. The data have been analysed in the MSc thesis of an Aberystwyth university student and are made available to the Shark Trust. This organisation runs the Great Eggcase Hunt, which began as a UK-wide citizen science project and has now become global. Providing records to the national Shark Trust database is important because this much larger sample will provide a more reliable distinction between real trends in shark and ray population sizes and bad weather events. The Penparcau surveyors found very large year-to-year variation in the number of egg cases. Shark and ray egg cases are usually rectangular, with tendrils at the corners. The eggs are laid among seaweed in relatively shallow water, and the tendrils are wrapped around the seaweed to hold the eggs in place. The gestation period is quite long – 10 to 12 months for the Nursehound in UK waters – and after this amount

FIGURE 4.5 A collection of shark and ray egg cases found on local beaches during one of the regular searches by the Penparcau citizen scientists. (Chloe Griffiths)

of time the empty cases can more easily be dislodged by storms. Thus stormy years will always result in a higher number of egg cases being washed up, but high winds could blow the dried egg cases off the beach, and these weather events are likely to mask population trends in such small surveys.

The recording of mermaid's purses is just one part of a much broader Penparcau village project called The Nature of our Village. This project is an inspiring community initiative that aims to create an inventory of all the village's wildlife, and to monitor trends and changes to it. It is almost a case of 'you name it and they monitor it' – from birds and flowering plants through a broad range of insect families to reptiles, molluscs, mammals, amphibians, sharks and rays. The project has a close relationship with its local environmental records centre, the West Wales Biodiversity Information Centre. The organisers from Penarcau village send in the records from all of its surveys, sometimes indirectly via the national databases for specific taxa, so that the data are available to national and international studies. The project also gets bacl spreadsheets and maps for the species that it has found – not just for its own data, but for all data held for Penparcau dating back to 1905.

The North Wales Wildlife Trust is running another citizen science project related to sharks and rays, but in this case there is an interesting twist on the survey technique undertaken by the volunteers – they do not need to leave home. It is called Sharks Inspiring Action and Research with Communities (SIARC), and it asks volunteers to examine underwater film footage taken by fixed cameras in the coastal waters of north Wales, and to identify and record any species that they see. The project is ongoing but, as an example, between May and September 2022 a total of 90 video cameras were set out in the Pen Llŷn a'r Sarnau (PLAS) Special Area of Conservation. This is a huge area to cover, consisting of 146,000 ha of ocean, so baited cameras were used to attract the sharks and were placed in likely spots – on sandbanks and reefs. The 146 hours of footage revealed quite a diversity of fauna, with 66 species identified in 90 hours of the footage, five of these being sharks or rays. Small-spotted Catshark *Scyliorhinus canicula* was the most frequent species. Interestingly, the Small-spotted Catshark did not appear in Penparcau's egg case surveys. In fact only two of SIARC's species did so, and the most common shark species in the Penparcau surveys, the Nursemaid, was not seen in the SIARC footage. Penparcau is just a few miles south of PLAS's southern boundary. These sightings and other records are being mapped together with data that characterise the physical environment by research scientists at Bangor University.

The computing power behind the SIARC project, which stores the images and video clips and delivers them to the volunteer searchers, is provided by Instant Wild, an initiative of the Zoological Society of London and the National Geographic Society. Instant Wild is the home of many local projects from around the world which use motion-triggered cameras to monitor animal populations and movements. For example, there are projects to monitor wolf populations in Croatian wilderness areas, jaguars and their prey in Mexico, great apes in the rainforests of Equatorial Guinea, the endangered Príncipe Thrush and, more prosaically, wildlife along London's railway network. Anyone with access to the Internet can do this, and clearly it can become addictive. Instant Wild runs a league table, and 'top spotters' have made tens of thousands of identifications and, in a few cases, hundreds of thousands of them.

Moving into the sea for real, many divers, free divers, fishers, paddle boarders and other water users are naturalists as well. They often provide important information to the scientific and conservation community or support its ongoing projects, sometimes even kickstarting them, and very occasionally disagreeing with the way forward. A case in point are the schemes to renew the kelp beds that were once extensive inshore along the West Sussex coastline. Two projects started at more or less the same time. One is proactive in nature, whereas the other is reactive. The kelp beds stretch along 40 km of the coast. They were formerly very diverse, a veritable forest, with abundant kelp providing a nursery and home for many fish, shellfish and other marine species. Anecdotally, local divers described the massive degradation of the kelp bed since the 1980s, and it has been estimated that 96% of the kelp had disappeared by 2000. The loss was attributed to the damage done by bottom trawling and by the great storm of 1987. Nothing could be done about the weather, but in 2021 a byelaw was introduced, with massive public support, to ban trawling along this stretch of Sussex coastline.

Steve Allnutt, one of the local free divers who had watched the decline of the kelp beds since he started diving and spear fishing with his grandfather off the coast of Shoreham as a young lad, was excited by the new byelaw but disappointed that no active measures were being taken to restore the kelp. He stepped in, sourcing free tanks on Facebook marketplace, increasing his shifts on his NHS day job and starting a crowdfunding page to pay for the rest of the equipment and the running costs. He gathered pieces of kelp from the Sussex beds, turned his garage into an experimental kelp farm and, using Instagram to garner help from scientists and kelp growers from North America, he eventually worked out how to get the kelp to grow and reproduce. The secret, he found, was to emulate the kelp's natural environment as closely as possible. This involved adjusting water flow (to mimic the upwelling and downwelling movement of currents and waves), water temperature and, very importantly, light quality and quantity. The light had to be switched between red-biased lamps giving just the right amount of photosynthetically active radiation during the day, and blue-biased lamps at night. Allnutt's project, known as the Sussex Seabed Restoration Project, continued to grow with more income from the crowdfunding scheme and a large grant from the FatFace Foundation, and there were recommendations from the Scottish Association for Marine Science and collaboration with Sussex University. With all of this help and support, Allnutt worked out how to harvest the sporophytes and grow them on substrates to be used as a vehicle for re-introduction. The substrates included oyster shells, gravel and biodegradable 'kelp lines'. Allnutt aims to experiment with these while restoring 3 ha of kelp forest. He planted kelp on the seabed in May 2023, and by September of that year it had tripled in size and he no longer needed coordinates to find it. This was the first demonstration that Sussex kelp could be restored in this way.

A second, much bigger project called the Sussex Kelp Recovery Project (SKRP) is taking a different approach. It might be characterised as the official project, being a partnership of seven local and national organisations, including two universities, the Zoological Society of London and the Sussex Wildlife Trust, with funding from the People's Postcode Lottery and other trusts and corporations. Its ethos is rewilding – that is, letting recovery occur naturally and observing what happens over an extended period of time, only intervening if problems become apparent. Citizen scientists are involved with SKRP's research programme, providing records of their observations. There is an app for recording the location and species of kelp seen by anyone who uses the beach or inshore waters, a programme through which underwater photographers can share their footage for SKRP to use to map the distribution of kelp and other species, and records from Seasearch, which is coordinated locally by the Sussex Wildlife Trust. Seasearch is yet another big science project. Its aim is to map the seabed found inshore around the whole of Britain and Ireland, and also to record any species encountered. It was created in the mid-1980s specifically for volunteer divers and snorkellers, who by 2021 had contributed almost 800,000 habitat and species records. Those records have directly helped the identification and creation of marine conservation zones.

There seems to be a quiet rivalry between the recovery project (SKRP) and the restoration project (Sussex Seabed Restoration Project). Neither mentions the other, and they have diametrically opposed views. The recovery project believes that nature should be given time to recover unassisted, whereas the restoration project believes that nature should be given a helping hand immediately. The restoration project is not introducing new species or even new genetic material, but rather it is bulking up kelp from specimens taken from the current degraded beds. This so-called reinforcement approach has been used elsewhere in conservation work – for example, when collecting eggs and hand-rearing Curlew chicks for release back into the wild. On the other hand, the rewilders think that active restoration may mask other problems that are preventing the full return of diversity, such as possible changes to the physical seafloor caused by trawling. The merits of the two approaches are currently an area of debate in the world of conservation, so what happens on the Sussex seabed with these two projects will be an interesting experiment in its own right.

On the west coast of Scotland, at Loch Craignish in Argyll, another citizen science project is aiming to restore lost biodiversity. This sea loch was once well endowed with seagrass meadows and oyster beds. By 2020 the seagrass meadows were much reduced, reflecting the UK trend (a 95% loss from coastal waters), and oysters had almost disappeared. Seagrasses build ecosystems, so they are keystone species. Their roots hold the sediments together and they provide a home for a large number of other species. Although they occupy only 0.4% of the area of Loch Craignish, they host 68% of its biodiversity. This is the first community-led seagrass restoration project in the UK. It has technical support from the scientists at Project Seagrass, a UK-based charity and, like the Sussex Seabed Restoration Project, it is proactive. The community volunteers have collected seed from the remaining plants in the loch and are trialling the best methods of introducing the seagrass to new areas – scattering seed, direct seed injection, planting seed in biodegradable hessian bags, transplanting sod and planting rhizomes. They have also built a seagrass nursery in order to determine what is the best medium for growing seedlings. The local aspect is important, as the experiments are testing the particular species that grows in the loch (there are four species that grow in the UK) in the particular environment of the loch. The volunteers and schoolchildren are monitoring the seagrass and the diversity that it brings. The Loch Craignish project, called Seawilding, is also restoring the beds of the native oyster by growing on spat (juvenile oysters) in cages in the loch and then simply scattering the live shells when they are big enough.

Staying with our watery theme and moving inland, there has been growing concern about the water quality of our rivers and its effect on the wildlife within them. Community citizen science has been very active in this area, too, perhaps not surprisingly given the affection that communities living near rivers have for them. In 2020 the River Wye turned pea green because of a massive algal bloom. Such blooms deplete the water of oxygen and block out the light, effectively suffocating the aquatic fauna. In one pollution incident on the Afon Llynfi, a tributary of the upper Wye, an estimated 45,000 fish died, and crustaceans were seen crawling out of the water. It is believed that the algal bloom was caused by enrichment of the river water with phosphate, primarily from agricultural runoff, and the finger has been pointed specifically at the waste from poultry farms (both intensive, if the chicken litter produced is used to fertilise fields in the catchment area, and free range). There are very many poultry farms throughout the Wye catchment. To give an idea of the size of the problem, consider Whittern Farm. Here chickens are not free range, and the farmer is able to collect their waste, incinerate it and thus both heat the chickens and produce phosphorus-rich ash, which is sold as fertilizer out of area to places where the soil is phosphorus poor. This single farm produces about 70,000 kg of ash annually. However, the problem is complex, and there are other potential causes of algal blooms, such as other types of agricultural runoff, human sewage, flooding events and global warming.

The local community decided to investigate, but quickly found that the amount of information available was very limited. Monitoring of water quality by the statutory authorities had been severely reduced by cutbacks, such that they only sampled infrequently and at a very limited number of sites in the Upper Wye catchment. There was insufficient data to demonstrate any causal effect between the chickens (or anything else) and the bloom. Consequently, the community decided to rectify this and sought volunteers to regularly collect and test water samples on their local patch of river. They started in 2021, and by 2023 there were over 200 volunteers, from all walks of life, testing samples up to twice weekly. The group has become a charity, Friends of the Upper Wye, and has merged with Friends of the Lugg. The River Lugg runs through Shropshire and is a major tributary of the Wye. Now the whole of the Upper Wye catchment is being frequently and systematically monitored. The monitoring is scientifically rigorous. Every volunteer follows a testing protocol designed by scientists at Cardiff University, each sample being analysed for electrical conductivity, temperature, phosphate levels, nitrate levels and turbidity (a measure of the quantity of particles, such as sediment, plankton or organic byproducts, in a body of water). They also record relevant site conditions, such as rainfall in the last 24 hours, water level and signs of pollution. The evidence that they have collected, together with research published in 2021 by Lancaster University using data from other sources (e.g. crop areas, fertiliser application rates, livestock excretion coefficients), has convinced the regulator, Natural Resources Wales, to concede that the blooms are not the result of high summer temperatures, as they had first concluded, and that poultry litter has caused pollution.

An important factor in the success of the Upper Wye monitoring project is the collaboration between amateurs and professionals, in this case between the community volunteers and scientists at Cardiff University. Virtually every big science project and a large proportion of small science projects will have this format. This is because, in order to make the results comparable and scientifically convincing, a protocol often needs to be developed and training offered. Sometimes the protocols must be very strict and the training rather intense. A good example of this is the study of the European Edible Dormouse *Glis glis* in Catalonia, Spain. The Dormouse Monitoring Programme was the first in Europe to use specific nest boxes, and has been running for 20 years, successfully providing biological and demographic information about this elusive animal at the southern limit of its distribution. The programme has two different protocols – the first one for less experienced volunteers, aimed at gathering qualitative presence–absence data, and the second one for experienced volunteers, aimed at gathering quantitative capture–recapture data. For the second protocol the animals must be handled in order to be ear-tagged and have their vital statistics recorded. A high level of skill is required for this, and the volunteers must attend a training course. The capture–recapture protocol provides a much richer data set, embracing juvenile recruitment, reproductive success, individual fitness and demography trends, and no doubt generates an intense personal experience for the handlers.

The existence of a small science project overseas is in no way remarkable. Citizen science is hugely popular around the world, and, just as in the UK, scratch the surface and you can find small science projects everywhere. The dormouse programme is one of four run by a research group based at the Natural Sciences Museum of Granollers. The other three are concerned with bats, small mammals and butterflies. Since monitoring began, roughly 1,000 volunteers have participated. In New Zealand, the Ornithological Society lists 75 regional projects. In the USA, as one might expect, there are many. There, where they seem to adhere more closely to the type of citizen science in which experimental design and data analysis are undertaken by professionals and data collection is done by volunteers, they have renamed it community science. A few examples from Utah will serve to illustrate the scene in North America. Concern about declining numbers of the American Kestrel *Falco sparverius* prompted the installation of hundreds of nest boxes to facilitate ringing and determine nesting success. During the spring nesting season, volunteer community scientists monitor the nest boxes and collect data such as number of eggs laid, morphological measurements, diet, physical health and fledging success. This is Hawkwatch International's Long Term Nest Monitoring Program. In the Wasatch Mountains, the Sageland Collaborative runs a wildlife watching programme in which volunteers set up and monitor trail cameras across the Wasatch throughout the spring and summer, and analyse the thousands of images taken. The aim is to enable better conservation planning in an area that is experiencing rapid development. At the time of writing, around 750 volunteers have helped to set up 2,500 camera stations which have captured more than 40,000 images of 28 species. Birds and Beavers is a project run by Tracy Aviary in Salt Lake City to explore the effect of North American Beaver *Castor canadensis* activity on bird habitat. It is centred on the Jordan River, which runs for 83 km between Utah Lake and the Great Salt Lake, through the most populous part of Utah. Volunteers, and indeed any members of the public, who observe beaver activity – that is, the gnawing of trees – along the river are asked to send in a photo together with the location of the tree being gnawed. Beaver activity has been reported along the entire length of the river, and the Tracy Aviary conservation team are accessing which of the more sensitive species of tree, such as cottonwood, need to be protected from further gnawing.

There are plenty of small science projects in developing countries, too, although funding can be a problem for these. The Cameroon Dragonfly Project employed local farmer Otto Mesumbe to collect, identify and rear out larvae. He became very skilled at this, and now works in California. Part of the funding for Otto came from the International Dragonfly Fund, a not-for-profit organisation based in Germany which uses the donations it receives to fund dragonfly projects around the world. The Charles Darwin Foundation based in the Galapagos Islands is another international not-for-profit organisation. It is dedicated to research into and conservation of the archipelago's biodiversity, and it has an interesting model by which it taps into the pool of affluent volunteers who are attracted by the location. These volunteers not only provide their time for free, but must also pay all of their costs, including travel, food and accommodation. Earthwatch, an organisation based in the USA, has taken this model a step further, and is rather like a broker between donors, researchers and volunteers, or perhaps it could be described as a citizen science travel agency, although its expeditions are serious projects – for instance spending one or two weeks investigating the diet of Killer Whale *Orcinus orca* in Iceland, collecting data on rhino locations, behaviour and habitat use in South Africa, or walking the beaches of Costa Rica at night to take measurements, collect skin samples and attach transmitters to Leatherback Turtle *Dermochelys coriacea*. A hefty participation fee pays for 70–80% of an expedition's costs, such as food and accommodation, equipment, training and research permits, and the volunteer must also cover their own travel costs.

Chapter 5

Gadgets

Citizen scientists have always used gadgets, adopting them just as soon as they have become useful, practicable and affordable. The standard gadgets in any particular field have remained almost the same for a century or more, with only two notable ones being abandoned – the vasculum and the shotgun. However, during the last 20 or 30 years the advent of modern electronics and computing has brought a whole host of new gadgets to expand the citizen scientist's horizons.

When Gilbert White was making his observations and conducting his experiments in the mid-eighteenth century, his essential gadget, and the only practical one available in his day, was a notebook and pencil. White took a different approach

FIGURE 5.1 Citizen scientists in action: a joint trip for members of the Cambrian Mountains Society and a local lichen group, Cennau Cymru, to look at lichens on Pumlumon. (Gill Foulkes)

to most botanists and zoologists of the time and focused on behaviour rather than identification, and on ecology rather than taxonomy. He did this by careful observation of the ordinary, together with painstaking record-keeping in his notebooks and journals, rather than by collecting the unusual and cutting it up. He was keen on recording his observations as they happened and would carry a field notebook for this purpose, whether it was to note the first song of a migrant bird or the actions of earthworms pulling leaves into their burrows as he watched them by candlelight. As old school as this approach might seem, the notebook and pencil are still one of the primary gadgets for natural historians, ecologists and experimenters.

White was a lone voice, and his more ecological approach was held in abeyance for almost a century. Darwin regarded him as one of his core influences, and it was with the publication of Darwin's works on evolution that the scientific world slowly moved away from its preoccupation with collecting. This led to the disappearance from botanical and zoological studies of two gadgets commonly employed by natural historians in the mid-nineteenth century – the vasculum and the shotgun. A vasculum was almost *de rigueur* for a botanist attending a field meeting in Darwin's day. It consisted of a cylindrical metal shoulder bag with a large door on the side to allow a whole plant to be transported from the field without damage. The inclusion of a damp cloth created a cool, moist atmosphere for the plant. Vascula gradually disappeared from the botanist's kit list and had fallen into disuse by the Second World War. The disappearance of the shotgun from the kit of ornithologists and entomologists was a result not only of the changing attitudes in response to Darwin's theory – namely the perception that humans are part of the natural world, not separate from it – but also of the availability of practical binoculars. Ornithologists no longer had to shoot birds to get a good look at them, and the same was true of large high-flying tropical butterflies that would formerly be brought down with dust shot.

Assisting the vision

A whole class of gadgets are designed to expand our senses. A pair of binoculars is just one of those that give our eyes a helping hand. Two optical gadgets became of practical use to citizen scientists in the mid to late nineteenth century, and in their various forms are still key tools for scientists today. These are the microscope and the telescope. A hand lens is a simple microscope, and a pair of binoculars consists of two telescopes joined together and working in tandem. Despite the astronomical work of Galileo and Johannes Kepler in the early 1600s using their own design of telescope, and Robert Hooke's publication of *Micrographia* in 1665 in which he illustrated the many objects he had studied using a microscope made by Christopher Cock, it was not until the mid-nineteenth century that practical, highly functional microscopes and telescopes became available. Both had required the manufacture of precision brass tubing, first achieved in the mid-seventeenth century, and of high-quality optical glass together with breakthroughs in optical design, which occurred during the nineteenth century.

Darwin owned a number of microscopes over his lifetime. The first one is thought to have been the exquisitely engineered, portable brass microscope which has been passed down through the Darwin family and is now in private ownership. It fits into a small case measuring 7.5 × 10 cm (3 × 4 inches) and has a live box accessory into which he would have been able to place samples of river and sea water for his studies of zoophytes. Darwin read his first paper on his microscopical work to the Plinian Society, a natural history club at the University of Edinburgh, in 1827, at the age of 18, when he was supposed to be studying medicine at the university. The microscope was bought for £11 16s, a tidy sum for any natural historian in those days, let alone a student. This may have been the same microscope that he used

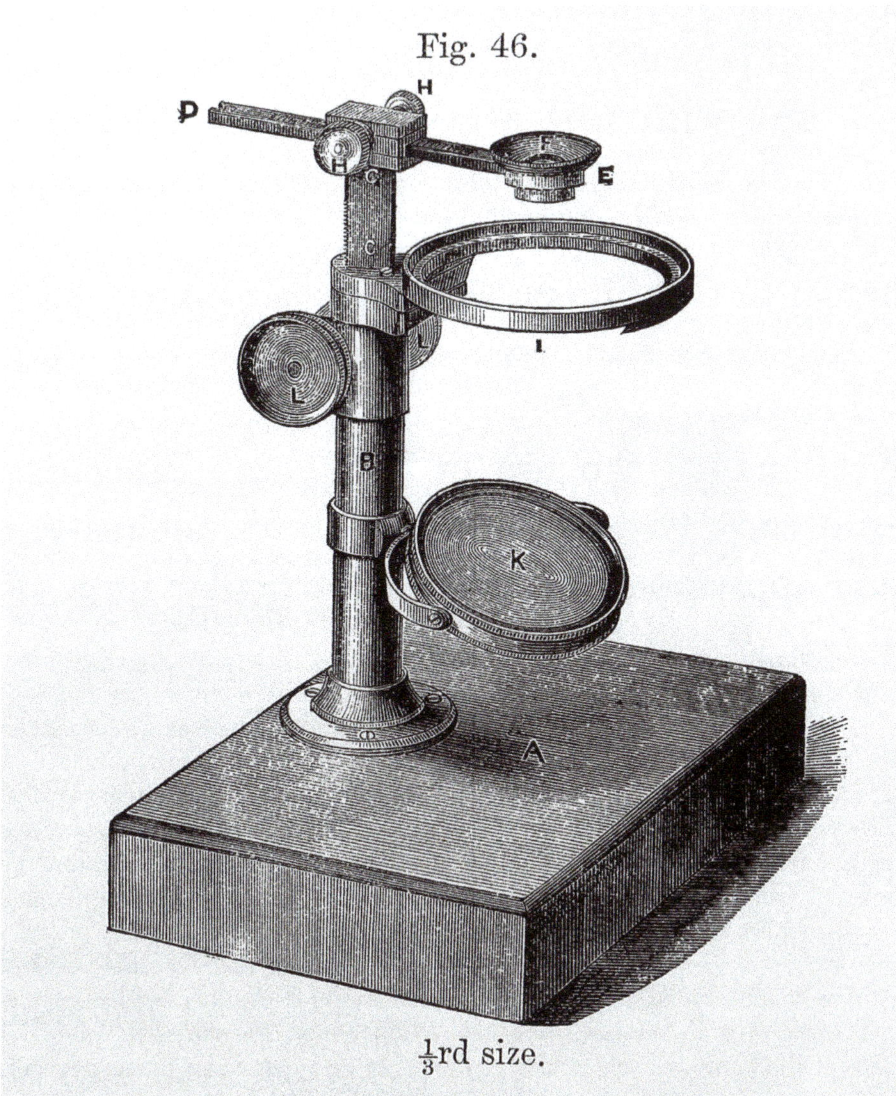

FIGURE 5.2 A simple, single-lens microscope of the type that Darwin took with him on the Beagle voyage. (Wellcome Library, London / CC-BY-4.0)

on the Beagle to study marine plankton. After the voyage, when he was working through all his specimens and developing his theory of evolution, he bought a more powerful, compound microscope. Darwin needed this to help with his studies on the classification of barnacles, because many barnacle subspecies can only be separated by viewing microscopic anatomical variations. He used his evolutionary ideas in the classification of barnacles, and he found that this classification supported his developing theory.

Darwin's compound microscope, which was made by James Smith, cost him £34 in 1847, so it too was an expensive instrument. Improvements in microscope design and an enormous increase in popularity, partly due to the description and naming of the cell nucleus by Robert Brown (1773–1858) in 1831, had meant that prices had dropped fivefold since Darwin acquired his first, portable microscope. Brown was another citizen scientist in his early years. Like Darwin he studied medicine at Edinburgh University, found it not to his taste, and turned to natural history, specifically botany, instead. He joined the army and was posted to Ireland, where he found himself with enough spare time to build a reputation as an expert botanist. His particular interest was in cryptograms – that is, lower plants such as mosses and liverworts. It was during this period that he took up microscopy, which would have greatly aided his studies of these plants. In 1800, aged 27, his growing reputation presented him with the opportunity to become a professional scientist. The burgeoning of microscopy among natural historians brought with it a surge of interest in all things small – algae, lichens, fungi, mosses, insects, spiders and so on. For many of these, and some higher plants, too, a microscope is needed to be able to distinguish between species. This is of course still true today, and many amateur scientists have a microscope or two among their gadgets. During the COVID-19 lockdown, for example, David Chelmick, 'stuck at home with my microscope', catalogued all of his Odonata larval material from Cameroon and produced an identification key to the genera.

At the lowest level of magnification, and an absolute essential for virtually every field naturalist except the ornithologist, is the hand lens or loupe. Even 'big' organisms occasionally need to be examined under a hand lens to confirm their identification. For instance, botanists use them to examine the hairs on forget-me-not, lady's-mantle and geranium plants to help to distinguish the various species. A hand lens allows one to see some of the delicate fine detail that nature has to offer, such as the opening and closing of the teeth around a moss capsule in response to the humidity of one's breath. Hand lenses commonly provide 10–20× magnification. Botanists looking at the hairs on higher plants will be happy with a 10× magnification, as will entomologists, but in their case because that magnification provides a larger field of view and better depth of field for viewing an insect that does not want to sit still. Bryologists and lichenologists looking at the finer detail of mosses and lichens, on the other hand, need a 20× magnification. The next step beyond this is the dissecting microscope, which provides a three-dimensional image, generally in the range of 20–60× magnification. This level of magnification allows the tiny structures that can answer very many identification puzzles to be viewed, such as insect genitalia, small mammal dentition, liverwort leaf insertion and fern sporangia.

Importantly, it is also needed to prepare slides for the next level of magnification provided by a compound microscope. These typically provide magnification in the range of 40–1,000× and allow one to see right into cells or view the fungal and algal partners in a lichen.

Some amateur naturalists devote themselves to the truly microscopic, studying whole organisms that can only be seen with a compound microscope, and the rather esoteric subject of amoebae provides some wonderful examples of this. Amoebae are single-celled protists that can alter their shape, typically by extending protoplasmic 'arms' called pseudopods or pseudopodia. They occur in a wide range of damp environments, are often aquatic, and can make up a significant proportion of the biomass of microbes in some habitats. Testate amoebae – that is, amoebae that have a hard shell and contain photosynthetic algae – can account for over 70% of the microbial biomass in peatland, which may mean that they are significant fixers of carbon here. As a group, amoebae are still relatively poorly understood and offer great scope for study by citizen scientists. For Dutch schoolteacher Ferry Siemensma they have been a lifelong avocation – he is a microscopist in the evenings, at weekends and on holidays. He happened to read about amoebae as a teenager and became determined to find one. He bought his first microscope at the age of 14, but it was not until he bought a second, more expensive model that used water immersion to improve resolution that he found his first amoeba. Now in retirement he has newer, more powerful models still, and on his website he has a large section describing his methods – for example, how to prepare testate amoeba shells from the debris collected with them. The real glory of his website is a huge gallery of images that he is gradually building up to form a survey of all amoebae worldwide. He has published more than 30 papers and books on the subject.

Another citizen scientist who is besotted with testate amoebae has gone one step further than would be possible for the majority of amateur microscopists. Eckhard Voelcker, who was a successful German software developer, sold his company in 2010 and, after several years of planning, used the proceeds to build a protistology laboratory in a room in his home. Then he took early retirement. The centrepiece of the laboratory is a scanning electron microscope (SEM) which can magnify samples by one million times or more. An SEM costs in the region of $50,000–1,000,000, plus the costs of installation, maintenance and all of the ancillary equipment required to prepare samples and process the machine's output. The results are amazing, and the website that Voelcker has created with his fellow amateur microscopist Steffen Clauß has galleries of exquisite images that show the complex structure of testate shells.

Optics in the form of binoculars and telescopes also enhance naturalists' vision by enlarging distant images. They are mostly used by ornithologists, although binoculars are also useful for entomologists to check out larger insects which are beyond their reach, such as butterflies and dragonflies. They are also of use to botanists for checking out plants in inaccessible places. The telescopes used by birders are called spotting scopes, and they are smaller, lighter, more easily manoeuvrable and of lower magnification, typically 25–60×, than the astronomical telescopes used for stargazing, which can provide magnification of 500× or more. However, the spotting scope needs a tripod and has a more limited field of view than binoculars, so its use is restricted to looking at relatively static distant objects. For example, it is commonly used for wader counts on estuaries when the birds are resting at high tide. The modern form of telescope had already been developed by the mid-eighteenth century, whereas usable binoculars took more than a century longer. The first binoculars combining Keplerian optics with a roof prism to provide a highly magnified, upright image were not made until the end of the nineteenth century. Modern Pechan-roof prism binoculars typically have 8× or 10× magnification, and are the ornithologist's workhorse gadget.

Spotting scopes and binoculars are essential tools for participants in most ornithological citizen science projects. A prime example would be the International Waterbird Census. This waterbird count has been running for more than 50 years, and many thousands of volunteers from more than 140 countries join in every year, making the census one of the largest and longest-running citizen science monitoring projects in the world. Wetlands are under threat everywhere and waterbirds are a good indicator of their health. With this in mind, the International Waterbird Census set out to monitor changes in the numbers and distribution of waterbird populations and identify internationally important wetlands. The census runs a strict protocol to ensure that the counts use the same methods at the same sites and the same time of year so that the totals are comparable across regions and years, even if they do not represent true population totals. The protocol states that binoculars are essential at all sites and that telescopes should be used on large sites where birds are frequently 500 m or more distant.

The advent of modern electronics has added only relatively modest improvements to the traditional optical aids, but has introduced one or two transformative devices. Hand lenses with built in LED lighting are now available, improving their usability in low light situations and offering other benefits that are not immediately obvious. For instance, they reduce the stress on bats that are being held in the hand for identification, as the handler no longer has to point the beam of their head torch at the animal. We now have access to gadgets that attach smartphone cameras to spotting scopes – so-called digiscoping. The combination can provide just as good (or even better) quality and magnification as the most expensive long lens camera, and is useful for providing verification of identification to citizen science endeavours such as eBird, or recording and reporting the colour of a bird's band ID.

Drones are one of the modern gadgets that have brought a significant extension to the naturalist's vision. In fact drones have been around since the First World War, but until the turn of the twenty-first century they were almost exclusively used by the military. In the early years of this century, several technological advances finally came together to facilitate the manufacture of the commercial drone as we know it today – advances in computer technology, miniaturisation of components such as motors and sensors, digital cameras, and lighter-weight and more efficient batteries. Together, these allowed for control of unstable but highly manoeuvrable quadcopters. One example of citizen scientists using drones to extend their vision comes from Western Australia. Here sea watchers recording the migration of the Humpback Whale *Megaptera novaeangliae*, Southern Right Whale *Eubalaena australis* and Blue Whale *Balaenoptera musculus* use drones to confirm the number of whales as well as their identity, size and age. Drones, or unmanned aerial systems as professional scientists like to call them, are increasingly being used in citizen science projects where the effort required exceeds the time and cost available to the purely professional and the study is being conducted in remote or inaccessible areas, or where disturbance must be kept to a minimum. Footage from drones flown by professionals may be analysed by amateurs – for example, counting marine iguanas in the Galapagos Islands or seals and sea lions in California. Alternatively, amateurs may fly the drones so that professionals can then analyse the footage – for instance, counting Koala *Phascolarctos cinereus* in remote Queensland, monitoring marine litter in Greece, or monitoring the dynamics of sandy beaches in response to storms and tides in Victoria, Australia.

Citizen science projects based on the use of drones have the same credibility issue with professional scientists as any other citizen science project, namely how valid the data are. When amateurs are flying the drones, training and strict protocols are required, so that the same start point, flight path, height, time interval and so on are used on every survey, and when amateurs are analysing the footage their output needs to be calibrated against a subset analysed by professionals. The seal and sea lion counting project in California was designed to test the veracity of counts by volunteers. It was found that the volunteer error level was relatively low, but not low enough, in the range of 5–9% in the breeding seasons when the species were gathered in separate groups, and the error was higher still during the summer when the species mixed. The citizen scientists tended to underestimate the number of animals. Of course the professional scientists must also validate the output of their own drone studies, and sometimes citizen science data are used to do this. For example, professionals using drones to study the relationship between waterbird breeding success and flooding levels on the Chobe River in Botswana compared their drone data with those from earlier land-based surveys by citizen scientists and found that both showed the same relationship.

The Koala study referred to earlier introduces us to another of the transformative devices created through modern electronics to expand our vision, because its drones were not recording images in the visible spectrum. They were using thermal imaging cameras which capture a picture of the infrared or heat landscape. All objects that are warmer than absolute zero emit infrared radiation to a greater or lesser extent, depending on their temperature. Warm-blooded animals stand out distinctly against their colder inert background, irrespective of the time of day, when viewed in the infrared spectrum. Thermal imaging devices have an infrared sensor with a special type of lens, often made from germanium, which is transparent to infrared, and software to convert the infrared image into a visible greyscale or colour-coded image. Those suitable for field use are usually in the form of a monocular, and they include a memory card to record the images. The lens is important because high-quality lenses add to the cost of an already expensive piece of kit, but they can make all the difference when one is trying to identify species. Identification to species level is not so easy with thermal images because they show the pattern of heat emission from an animal, which will not necessarily match its visible shape and certainly will not match its colour pattern. Experience of a species' heat signature and resting or moving behaviour is helpful here, or night vision devices can be used. These devices should not be confused with thermal imaging devices. They are optical devices that work by using electronics to amplify the visible light that they capture. Night vision devices are better for identification, but they are not very good at finding the animal in the first place, whereas animals are easy to spot with thermal image monoculars, but not so easy to identify. The Koala project solved the identification problem by using artificial intelligence (AI) to scan the images that the drones had captured, looking for the Koala's thermal signature and resting position, and it may be that in the future AI can be developed to offer real-time identification assistance to users of thermal imagers.

Thermal imaging has been especially transformative for ornithology. Citizen scientists in the West Midlands Ringing Group have done a lot of the pioneering work in the UK to discover just how useful they are. The group managed first to borrow a monocular and then to raise around £2,500 to purchase one of their own. The group is mainly interested in farmland birds. They were prompted to start thermal imaging when they realised the difficulty of spotting Lapwing *Vanellus vanellus* chicks, particularly for inexperienced ringers. During a 30-minute daytime session two experienced unaided ringers were able to find just one chick, which a trainee ringer who was with them could not see at all. A week later, at the same site and over a similar time period but this time using a thermal imager, they counted 29 chicks and also found

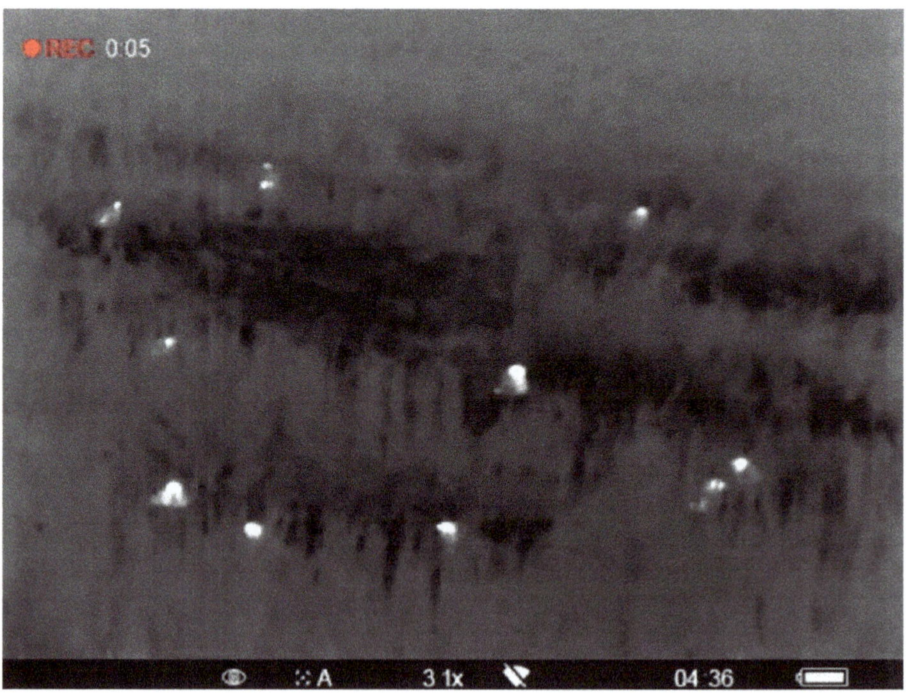

FIGURE 5.3 Surveying Skylark *Alauda arvensis* at night using a thermal imaging camera. It is clear that this gadget will improve the accuracy of such counts and significantly reduce disturbance compared with the more traditional method of quartering a field and using a torch. (West Midlands Ringing Group/CC-BY-SA-4.0)

nests with eggs or newly hatched chicks. Clearly the thermal imager gives a huge improvement in the accuracy of the estimates that the group submit to the BTO Nest Record scheme, and saves a great deal of volunteer time and effort.

Thermal imagers are equally effective at night, and perhaps more so, where a major benefit is reducing disturbance and consequent stress on the birds. Rather than tramping up and down fields with torches, surveys of roosting birds such as Skylark *Alauda arvensis* can be done from the edge of a field. Night-time surveys have turned up flocks of birds that local farmers thought had long since disappeared, such as Golden Plover *Pluvialis apricaria*. The thermal imager also makes all forms of netting for the purpose of ringing easier. Examples include Jack Snipe *Lymnocryptes minimus* during the day, because it is possible to approach it from behind, or Woodcock *Scolopax rusticola* at night, because it is possible to dazzle the bird by pointing a torch beam directly at it. It also helps to reduce the length of time for which a bird is trapped in a mist net at night, because it is possible to monitor from a distance. Finally, the device aids some breeding bird surveys. It can detect whether the burrow of Sand Martin *Riparia riparia* is occupied, or whether a nest box is occupied. Studying skylarks with a thermal imager has shown how important arable stubble is for wintering birds, and that they will not roost in cover crops once these have grown to 8–10 cm in height. Thus the device has helped to improve land management advice by showing that zero till is better than ploughing for overwintering birds.

The third transformative device to expand the naturalist's vision is the trail camera, the modern version of which is basically a digital camera triggered by an infrared motion sensor. Trail cameras can sometimes have a wireless connection to allow them to be managed remotely and their images to be retrieved without having to visit the camera. The first trail cameras, introduced in the late 1980s, were just film cameras linked to motion sensors. They were bulky, expensive, impractical and time consuming. What was really needed was a digital camera, and these had to await the development of miniature semiconductor image sensors and memory cards, which did not come together in a suitable form until the late 1980s. The first fully digital camera was not introduced until 1988, but it could only store 10 images of low resolution. Thus it was that the first major step forward for the trail camera came with the ready availability of higher-capacity point-and-shoot digital cameras in the mid-2000s. Further improvements were required in energy use and battery life to result in a really practical device, but by 2010 trail cameras were in wide use in citizen science projects.

Citizen science projects using trail cameras first appeared in North America, which is not too surprising because one of the main drivers for the production of consumer-level devices was recreational hunting. As well as the low price of the devices, camera trapping is ideal for surveying animals, particularly mammals, on a landscape scale by volunteers with limited experience. Prior to the introduction of trail cameras, surveys were restricted to smaller areas because of the need for experts who could read tracks and signs, as mammals are usually shy, often nocturnal and do not vocalise much. Not only are volunteers keen to take part in camera trapping because they find it a rewarding activity, but also their observations are automatically vouched for by the image. A certain amount of associated information or metadata, such as the date, time and location, is also automatically recorded and linked to the image. Numerous citizen science projects of all sizes have been and are being run to look at the distribution of mammals using trail cameras. These range from very small studies, such as the Fellfoot Forward project which has four camera traps to loan to interested citizens in their area of the North Pennines, to very large ones, such as MammalWeb in the UK and the Wasatch Wildlife Watch Project in Utah. Between 2013 and 2022, MammalWeb has monitored wild mammals for the equivalent of 340 camera trap years with over 180,000 detections. Much of the camera deployment and image analysis has been done by citizen scientists.

Camera traps can be used to do more than record a species distribution. It is possible to use them to study behaviour and population structure, too. This needs a more structured design and protocols, and must also attract and coordinate the many participants required. eMammal is a web-based system designed to support such studies. It was one of the earliest camera-trapping, crowd-sourced initiatives and was set up in 2012 with funding from the US Federal Government. During its first year, 85 volunteers set camera traps in nearly 700 sites and collected 25,000 observations. These suggested that the recreational use of land might alter the structure of the animal communities that lived there. Certain animals, such as White-tailed Deer *Odocoileus virginianus* and Black Bear *Ursus americanus*, were detected more frequently in areas that were not hunted and also more frequently off-trail than on-trail. Another early eMammal study placed camera traps at over 2,000 sites to compare the use of protected and urban areas by the Domestic Cat *Felis catus*. It seems that cats and Coyote *Canis latrans* do not get along, because in areas where coyotes are common, in protected areas and urban fringes, cats are rare. eMammal is still operating and its use has spread to projects worldwide. MammalWeb is a similar system, in this case funded by the EU and covering the UK. At the time of writing, MammalWeb is hosting over 40 citizen science projects.

Camera traps are generally optimised for large and medium-sized animals, so small mammals can be a challenge. A distant view of mice at night, when they are most active, is likely to show just the two bright spots of their eyes, whereas a close-up can be blurred or washed out because the infra-red flash is too bright. At the same time the small mammal must get close to the camera if it is to be identified. To overcome these conflicting problems, DIY modifications have been developed for small mammal studies. In traps for mice, voles and shrews the camera is strapped to one end of a baited box, and the other end is left open. This brings the animals close to the camera, which has a close-focus lens fixed in front of its built-in lens and a layer of paper to reduce the brightness of the flash. These modifications were tried out in the Flow Country of northern Scotland where 80% of the images that they captured were of mice, voles or shrews. The by-catch consisted of ground beetles, robins and Weasel *Mustela nivalis*. On the subject of weasels, researchers in the Netherlands have developed a modification specifically for surveying mustelids, particularly Stoat *Mustela erminea* and weasels. This trap exploits the particular liking that mustelids have for drainpipes. It is similar to the mouse camera trap but bigger, the open end is replaced by a piece of drainpipe, and the camera is protected from

FIGURE 5.4 A camera trap modified for recording small mammals. (Littlewood *et al.*, 2021: https://doi.org/10.1007/s10344-020-01449-z)

the mustelid by a piece of wire screening. The Vincent Wildlife Trust has been using mustelid camera traps to look at the population status of stoats and weasels in the UK, and citizen scientists are helping to review the video footage via MammalWeb.

Assisting the hearing

Gadgets designed to give our ears a helping hand did not become available to the naturalist until quite recently. Field recording and playback devices first became a practical proposition in the 1960s with the development of portable audio recording equipment using magnetic tape. However, the real advance for citizen scientists came in the 1990s with further miniaturisation and development of digital recording on to solid-state storage. At the time of writing it is possible to buy even more advanced, pocket-sized devices, such as the AudioMoth, for as little as £65. The classic citizen science projects that make use of audio devices are bat surveys. Bats emit ultrasonic sounds that are mostly beyond the range of human hearing. These high-frequency echolocation and social calls are definitive and can be used for identification.

The earliest and longest-running volunteer-based regional bat monitoring schemes normally use affordable and relatively simple devices called tunable heterodyne detectors. These work by in effect subtracting the bat's audio frequency from a constant tunable frequency and playing back the remainder, which will be audible to the human ear. This means that the audio will be incomplete, clipped when the two frequencies are the same, so the identification needs to be done in real time because the recording is not of good enough quality for later analysis by computer. These limitations mean that it is difficult to distinguish between the calls of many species. Nevertheless, tunable heterodyne detectors do allow confident identification of some species by using a combination of call repetition rate, rhythm, tone and peak frequency. Consequently, they have been used successfully for more than 25 years by volunteers taking part in the UK's National Bat Monitoring Programme (NBMP). Five species are monitored in the NBMP, which uses a combination of field surveys with detectors and roost counts to monitor ten species in total. In the 15-year period to 2012 around 19,000 hours of volunteer time were spent on the programme, with an estimated value in excess of £250,000. With training and strict protocols the surveys were found to produce robust data capable of being used to determine whether the suspected decline in numbers was sufficient to justify red listing the investigated species. Surprisingly, however, the data showed that in nine of the ten species the numbers were stable or increasing over the study period. That is not to say that the perceived steep decline in living memory is incorrect, but because there are no earlier national data this first period of the NBMP must form the de facto baseline.

More sophisticated, and consequently more expensive, handheld detectors intended for real-time use by active bat observers are readily available. These range from heterodyne detectors with auto-tuning to time division detectors, which slow down a brief snapshot of the full spectrum of sound so that it becomes audible, or a combination of the two. Some detectors also record the acoustic clips for later playback. A clever innovation is an attachment that turns a smartphone into one of these sophisticated devices for the cost of one of the lesser detectors. For example, the Echo Meter Touch works with smartphones running the Android operating system and provides auto-tuned heterodyne and real-time, time division listening, together with an app that displays a real-time spectrogram of the clip and shows the two species most likely to have made the sound.

There are obviously limitations to such 'snapshot' surveys where only a few species are surveyed over a relatively short period of time. Ideally, comprehensive surveys would consider all species over longer periods, preferably all night. This is increasingly possible for citizen science projects with detectors such as the Audiomoth, which records full-spectrum sound from audible to ultrasonic frequencies on to a microSD memory card. This device is small – about the size of a stack of a dozen credit cards – and power-efficient enough to be just attached to a tree and left overnight. Such long recordings with many inaudible ultrasonic calls need to be processed by fairly sophisticated computer programs, which are freely available via various websites, such as the BTO's Acoustic Pipeline.

In projects based on such passive acoustic recorders, citizen scientists are usually responsible for managing the deployment of the monitors, uploading the recordings for analysis and in some cases, for the more experienced, doing the analysis. For instance, a Catalonian project that is studying the Edible Dormouse is also monitoring bat populations in part by using passive acoustic recorders. Fixed stations are sampled either for seven consecutive nights in the summer or for one night in each season. The passive acoustic recorders have recorded 15 of the 25 species of bat found by the programme as a whole, and of the 30 species known to occur in the area. These data have been used to foster ecologically friendly agricultural practices.

A similar project, called Ryevitalise, has been running for a few seasons on the North York Moors. Again citizen scientists deploy the detectors and upload the recordings, in this case to the BTO's Acoustic Pipeline. In 2022 they managed 413 nights of recording over 100 sites during the spring and summer, collecting more than 250,000 clips of bat calls. The data show not just species distribution but also activity patterns, and bats are not the only animals recorded. The 2022 survey also detected the calls of shrews, rats, a bush cricket and a couple of moth species. Of course bats are not the only animals that vocalise, and citizen scientists have been involved in bio-acoustic projects for a range of species. Two projects of note have looked at mass participation by people using their smartphones together with a custom-built app. The New Forest Cicada project involved 3,000 participants using an app built by a researcher at Southampton University to search for the UK's only native cicada, *Cicadetta montana*. The cicada, thought to be extinct in the UK, has a similar song to two more common bush crickets. The participants were asked to make recordings, and if the app considered that any of these were possibly of the cicada, they were uploaded to a central computer for further analysis. In total, 11,000 reports were uploaded over the two seasons for which the project ran, but unfortunately none of them were of a New Forest Cicada. The second project, based in Berlin, was designed to test the quality and quantity of recordings of Nightingale *Luscinia megarhynchos* made by amateurs using their smartphones and the Naturblick mobile app by comparing them with those made by experts using professional equipment. The upshot was that the citizen scientists were just as capable as the experts, and any difference was due to the quality of the equipment.

Cetaceans are famously vocal. Submersible passive acoustic recorders have been developed for studying their sounds and making identifications for species surveys. These devices tend to be rather expensive, around £2,000 at the time of writing, and are mostly used by professionals, but citizen scientists are sometimes involved. For instance, as part of its study of the locally threatened population of Bottlenose Dolphin *Tursiops truncatus* the Cornwall Wildlife Trust ran a project to test a new version of the C-POD passive acoustic monitor while also monitoring the local populations of various dolphin species. Citizen scientists were able to participate in two ways. First, volunteers deployed the C-POD, and secondly, land-based volunteers doing their standard cetacean watch added to their protocol by also noting birds feeding or diving near the C-POD, and the proximity and direction of travel of cetaceans in relation to it. Both of these additional observations aid interpretation of the cetaceans' vocalisation. By matching the sightings data with the acoustic records the manufacturer of the C-POD (a Cornish company) was able to gauge its effectiveness and to use the minor issues encountered, such as the problem of sediment movement noise, to refine its development.

It has been suggested that passive acoustic recorders are better for recording the presence of wildlife than trail cameras. It is argued that since they are not particularly directional, they can record any vocalising animals within the whole hemisphere of their range, whereas trail cameras need their subjects to move through the camera's field of view. Also, trail cameras are not so good at capturing images of small organisms. Of course, many animals, especially mammals, do not vocalise when they are moving about, so the two types of device are actually complementary. In fact the combination of devices is game-changing in certain situations – for example, using night-vision binoculars to identify animals found by thermal imaging, or using thermal imaging to count birds migrating at night that have been identified by acoustic monitoring. Participants can also play back sound to birds found by thermal imaging to cover the approach of bird ringers.

A neat demonstration of the benefit of using multiple gadgets in citizen science comes from a bird translocation project in the South Island of New Zealand. Here the local community in the Flora Stream area of Kahurangi National Park teamed up with the Department of Conservation to reintroduce the Great Spotted Kiwi or Roroa *Apteryx haastii* into an area where it had become extinct, primarily due to predation by introduced stoats and weasels. In order to show whether the reintroduction was successful, the new population had to be monitored for long enough to demonstrate growth. This is tricky in the case of the Roroa because it has large territories, it is nocturnal, it has naturally low productivity and it is very sensitive to disturbance – just walking past the bird can cause it to abandon a nest. The project solved these issues by running the monitoring for eight years, fitting radio transmitters to the adults to monitor them and using two trail cameras, placed with great care, to monitor each of the nests. Neither of these methods was without difficulties. The study area was large. It covered 10,000 ha (100 km^2, a hectad) of mostly southern beech forest over an altitude range of 700–1,500 m, into which only 44 birds were released. Finding the adult birds by radio tracking during the day was relatively straightforward once their roosts had been located in the first place, but tracking their activity at night required triangulation of a moving bird from three or four points every 20 minutes. Volunteers had to spend thousands of hours tracking the kiwis. Because of the risk of nest abandonment, the trail cameras were placed no less than 4 m from the nest entrance, which meant that they were not always triggered in time to capture video of the small chicks and potential predators. Furthermore, some nests had more than one entrance. Despite these difficulties the study was able to demonstrate that the translocation had been successful. The target was a 2% annual population growth rate, whereas the measured rate was 3.4%, albeit with fairly broad margins of error. Half of the chick deaths recorded were due to predation, and the culprits were found to be stoats or weasels. The native Weka *Gallirallus australis* was often a nuisance, and a small proportion of incubation failures were attributed to them, but none of the chick deaths. The project's success would not have been achieved without the other major contribution of the Flora community, namely an extensive and continuous effort to eradicate non-native mustelids. Around 1,000 traps are set in the area and maintained by the community. Monitoring has continued by using a third, less labour-intensive technology – acoustic monitoring of Roroa call rates.

Bringing organism and scientist together

We do not need to expand our senses if we bring the animals to us, and there have long been devices designed to humanely trap or entice animals to come close for observation. An early, rather bizarre method of using entrapped animals to study behaviour was tried by the German zoologist couple Oskar and Magdalena Heinroth. They both worked professionally in zoology, Oskar being director of the Berlin Aquarium for 30 years. They were also keen ornithologists, interested in bird behaviour in particular. Oscar actually proposed to Magdalena with the gift of a Blackcap *Sylvia atricapilla*. They ran a project to study bird behaviour by raising chicks and keeping free-flying adults in their apartment. Over the years they brought up over 250 European species, with some interesting consequences for their apartment. For example, the Tawny Owl *Strix aluco* would pounce on the cuckoo on the hour as it appeared out of the cuckoo clock, and a Lesser Spotted Woodpecker *Dendrocopos minor* drilled its nest hole into their cupboard. It is testament to his dedication to the study that Oskar was prepared to sleep wearing an oxygen mask because he became allergic to the birds' dust.

Thankfully, today the normal methodology is to work in the natural environment, only using traps for a brief encounter with the animal. Traps are generally used for smaller animals, and entomologists are perhaps the most common users of them, with their nets, pooters, pitfall traps and moth traps. The butterfly net has been around for many centuries in more or less the same form – a voluminous bag of light material held open by a stiff hoop, with a short handle. In fact British entomologists used a version called a clap net until the mid-nineteenth century. This awkward-looking device had two nets, one attached to either arm of a scissor mechanism, the idea being to approach the insect with the scissors open and then to snap them shut to trap the insect inside. However, once the British naturalists learned how to flip the net to stop the insect escaping, the continental-style single net took over and the clap net disappeared from use. Today there are multiple designs to cater for various levels of user and different kinds of insect. Butterfly nets, used for flying insects, have a net made of lighter material and a lighter, less robust hoop so that they are more manoeuvrable and less likely to damage the insect's delicate wings. Sweep nets, used for a wider range of insects such as flies, beetles and grasshoppers that are settled on vegetation, are made of heavier material and have more robust hoops, to help to prevent damage as the net is swept through the vegetation. Pooters are used to suck small insects into a collecting jar or vial. They can be used in combination with a sweep net or to take an insect directly off vegetation or a surface where one would not use a net, such as dung. Basically they consist of two tubes that feed into a sealed container. One tube is pointed at the insect while suction is applied to the other tube, either by using a puffer ball or by the entomologist sucking. It is probably safer to use a puffer ball, to avoid any risk of sucking the insect or the dung into one's mouth! The pitfall trap is simply an empty jar, yoghurt container or similar container that is sunk into the ground to the level of the soil surface. Ground-dwelling insects fall into the trap and cannot escape until the entomologist inspects it. The efficiency of pitfall traps can be significantly improved by connecting two of them together with a barrier, such as a length of aluminium angle. Insects that walk into the barrier tend to turn and follow the barrier until they walk into one of the traps.

Moth traps revolutionised the study of night-flying insects, and they are popular enough for their enthusiasts to have become known as moth-ers, and their hobby as mothing. The traps work by making use of the moths' propensity – still not fully understood – to fly towards a source of light at night. Indeed, some moth-ers started out by making morning visits to public places where the lights had been left on all night, such as telephone boxes and toilets. Moth trap efficiency and practicality have improved in line with the development and commercialisation of lighting – oil-powered hurricane lamps in the mid-nineteenth century, incandescent lamps at the end of that century, mercury-vapour lamps in the mid-twentieth century, actinic fluorescent lights in the late twentieth century and LEDs in the early years of this century. The early traps simply consisted of a lamp held against a white surface, typically a sheet or a wall, whereas modern traps are fully integrated devices in which a bulb is suspended over a box with vanes to deflect the spiralling moths away from the light and into the box. Traditionally, old egg boxes are placed at the bottom of the trap for the moths to rest in. A wide range of moth traps are available, designed for use in different situations, with various bulb types, wattages and power sources. In general, brighter bulbs will attract more moths.

In the UK, two main centres curate moth records, and citizen scientists contribute to both of them. The Rothamsted Insect Survey (RIS), which started in 1964, now has 80 traps, all of them 200 W clear tungsten filament bulbs, situated across the country and mostly run by volunteers. Since its inception the RIS has trapped 1,500 different species of moth. The National Moth Recording Scheme (NMRS) started much later, in 2007, gathering records from thousands of moth-ers all over the country, and running a wide variety of traps. Its database holds in excess of 34 million records. Professional scientists have used data from both databases to map moth distribution and population trends over time. In Scotland, for example, a study that used the RIS data to analyse trends in abundance and the NMRS data to examine trends in distribution found apparently conflicting results, namely a long-term decline in abundance but an expansion of the range of night-flying moths. It is suggested that the warming summer climate is responsible for the expanding distribution, and a combination of habitat degradation and warmer, wetter winters is responsible for the fall in numbers.

Relatively efficient live trapping of small mammals in the UK resulted from the development of the Longworth trap in the late 1940s, when population studies and control of small rodents became important for the post-war agricultural economy. Live trapping studies of small mammals had started in the 1930s, but with much less sophisticated equipment. The Longworth trap is still very widely used today. It has two components – a nest box that is baited and filled with bedding material, and an entrance tunnel with a trap door which is closed when the animal stands on a trip plate at the nest end of the tunnel. The sensitivity of the trip wire can be adjusted for animals of different weights, and the nest boxes are available with or without a shrew hole to allow these particularly vulnerable animals to leave, as in the UK only licensed individuals are allowed to handle shrews. There are various other designs of live traps – the Sherman trap, for instance, is a single, collapsible aluminium box, again with a trap door – as well as non-trapping designs, such as footprint tunnels, hair tubes and faecal tubes. Each has its advantages and disadvantages. The non-trapping designs are less expensive to operate, but identification from the traces left by the animals that pass through them is difficult or expensive if DNA analysis is required. Perhaps the best solution for simple presence/absence surveys is the modified camera trap, which collects images over time and date stamps of all the animals that enter, and they are not stressed in any way. Conversely, live traps capture only the first animal that comes along, so long as it is not too light to prevent the trap from triggering, or too big to prevent the door from clicking shut. Some stress to the animal is involved, but live traps do allow the recording of sex and weight, and the marking of animals for capture–recapture abundance studies. One advantage of live trapping for citizen scientists is that they can get up close and personal with small mammals which they would not otherwise often get to see. This can greatly increase dedication to projects that can run for long periods and at unsocial hours.

Volunteers have been taking part in Mammal Society live trapping studies since the early 1980s, and these continue today at the national and local level. Since Longworth traps are quite expensive and many are needed for a thorough survey, the society loans traps. It can also provide training. One of its first surveys involving volunteers in live trapping was the National Woodland Small Rodent Survey, which ran from 1982 to 1995. It involved Wood Mouse *Apodemus sylvaticus* and Bank Vole *Myodes glareolus* trapping in spring and late autumn in 30 woodlands using nearly 3,000 traps. This long-term study provided some interesting information on the dynamics of the populations of these small mammals, particularly for the Wood Mouse, which showed a synchrony of fluctuation in numbers in spring between the 30 sites. The Bank Vole population did not vary so much, so any synchrony was difficult to detect. For the Wood Mouse population, the fluctuations appeared to be due to a combination of seed masting (a good seed crop in autumn was correlated with higher spring numbers), weather (one particularly harsh winter caused a crash in the numbers of both species) and density dependence. Over the whole period the populations of both small mammals remained fairly stable.

Some studies of animal life history and population dynamics can be done without trapping, instead using various enticements to bring the animals to the observer. Entomologists have long made use of insects' olfactory desires when studying them. Before lighting was good enough to make effective traps, moth-ers had discovered that many species of moth are attracted to sweet-smelling substances. The concoctions that they prepared are still used today for sugaring, in which a thick mixture of molasses, brown sugar and ale is painted on to a wall or tree trunk, and wine roping, in which a length of rope or a strip of fabric is dipped into a mixture of red wine and sugar and draped over a branch. More modern equivalents are less messy. It has recently been shown that amyl acetate, a simple organic compound with a fruity smell, can increase by one-third the number of moths attracted to a light trap. Pheromone lures – that is, chemical cocktails mixed so as to mimic the pheromones released by female insects to attract males – are also now commercially available. Admittedly many of these pheromone lures target pest species, having been designed to give early warning of potential infestations, but they are not always absolutely specific, and non-target insects can also be lured. Increasingly, pheromone lures are becoming available for non-pest species, some of which are of conservation interest and may otherwise be difficult to monitor, such as the day-flying clearwing moths. Pheromone lures are very effective, fast-acting and have a long life, so when they are used for non-pest species this needs to be done responsibly, especially if they are attracting males of scarce species.

Birds and small mammals can be encouraged to the scientific observer by feeding stations and nest boxes. John Dovaston (1782–1854), another early citizen scientist, is credited as being the first to create something akin to the modern bird feeder – a device he called the ornithotrophe. He was also an early experimenter with nest boxes. Dovaston was fortunate that his self-educated father had been so singularly successful that Dovaston himself could spend most of his life in rural retirement focusing on his two avocations – poetry and nature. He was a remarkably innovative naturalist. As well as creating his bird feeder and nest boxes, he was the first to demonstrate by marking them that robins are territorial, having attracted them using his ornithotrophe. This led him to become a pioneer in bird ringing. He used neck rings made of the strings from his cello to ring swallows, and amazingly some of his ringed birds survived and returned the following year. He was also one of the few naturalists to use a spyglass for watching birds, when most were still using a gun. He even experimented with growing Mistletoe *Viscum album* on various tree species, and was successful with a dozen or so. He also attempted, but failed, to use musical notation to record birdsong. Despite these really remarkable efforts, he seems to have regarded his nature studies as rather trivial compared with his poetry, and he was rather slack about publishing any of his work (when he did, he would sometimes use a pen name – Von Osdat – an anagram of his surname).

Today nest boxes tailored to different types of animal, and even individual species, are readily available commercially. They include nest boxes for small and large birds, swift boxes and bricks, sparrow terraces, and boxes for robins, house martins, swallows, various owls, dippers and wagtails, hedgehogs, squirrels, dormice, pine martens and bats, among others. There is even a heated bat box for a maternity roost. Boxes can be fitted with webcams and can be fabricated from a range of materials. An interesting variant has been used by a bird translocation project that has relied heavily on volunteer effort, not least from the local community, on the South Island of New Zealand. The project's aim is to create a safety-net colony for the endangered, endemic Hutton's Shearwater *Puffinus huttoni*, which is known from only two other colonies, both high in the Seaward Kaikoura mountain range on the north-east coast of South Island. These colonies are vulnerable to feral pigs and cats, and to landslides caused by earth tremors. The earthquake in 2016 destroyed about 12% of one of the colonies at the height of the breeding season, killing as many as 120,000 birds. The special wooden nest boxes created for the project are sunk into the ground and connected to the surface by a length of 12 cm drainpipe tunnel.

They have removable lids to enable access to the chicks. The project, which has been running since 2006, has become one of the largest seabird translocation projects worldwide. To date, more than 500 pre-fledgling chicks have been translocated from the mountainous original colonies to the new coastal colony.

The chicks brought down from the mountain are raised in the nest boxes and fed on sardine smoothies administered largely by the many local community volunteers who also played a big part in the monitoring of the movement of birds. The chicks stay at the site for about four weeks, which appears to be long enough for them to become imprinted on their new home, as very few that return from sea as adults after three to six years have gone back to their original colony, and none of the chicks that have been born in the new colony have ever been seen as adults in the mountain colonies. By the 2020–2021 season the new colony had produced 112 fledglings, and had grown steadily to a population of about 80 adults. This is seen as a great success, but it will be many years before the colony becomes full, since there is space within the predator-proof fence for 4,000–10,000 birds, depending on burrow density.

In the early days, monitoring was done in daylight hours by looking for displacement of metal pins placed at each end of the nest-box tunnel. It was assumed that this indicated when chicks left the box or parents visited, as in theory the birds would knock over the pins. Later the birds were fitted with electronic Passive Integrated Transponder (PIT) tags, and these showed that the pin displacement method was most unreliable. It turned out that the pins were sometimes knocked over by Song Thrushes *Turdus philomelos* looking for snails, and by chicks from different nest boxes. Chicks were found to habitually wander at night, and would visit up to four other boxes. The boxes were monitored weekly, and only during the day, because the Department of Conservation feared disturbance of the chicks if they were inspected at night. The PIT tags showed that this method missed about 50% of the adult bird movements.

The PIT tag is another piece of technology that has found its way into the scientific toolkit and is not beyond the price range of the citizen scientist for certain applications. It is based on radio-frequency identification, more commonly known as RFID. Today RFID solutions are ubiquitous, and because the tags that they use are so tiny they can be inserted into, or attached to, anything for which unique identification is needed – pets, passports, packages and so on. Ornithologists use a PIT version combined with a microchip. These are usually incorporated into a plastic band that can be as small as 2.5 × 8.5 mm and can weigh as little as 0.1 g. An RFID solution has two basic components – the tag and a tag reader. When the tag enters the electromagnetic field produced by the reader, it is induced to transmit its unique identification number, which is then picked up and logged by the reader. The tag does not need its own battery, so its lifespan is theoretically unlimited, although general wear and tear is likely to reduce this to a few years. One problem is that the small size of the tag and the frequency available for bird bands means that the read range is quite small (less than 1 m). For this reason the readers need to be positioned where the birds can be guaranteed to gather, making the combination of a PIT tag and a feeding station or nest box ideal.

PIT tag bands are being used by the volunteers at Treswell Wood (whom we met on page 74 in the chapter on small science) to glean a better understanding of the life of the Marsh Tit *Poecile palustris*. The Marsh Tit has suffered a severe and continuing decline in numbers in Britain. Between 1967 and 2020 the UK population decreased by 80%, so in 2022 this species was red listed. Long-term monitoring by ringing and Common Bird Census surveys at Treswell has shown a slightly different pattern. It too has suffered big fluctuations in its population of Marsh Tits. In fact the bird became extinct in the wood in the harsh winter of 1978–1979, but it recovered due to the influx of immigrating birds and has remained stable ever since, bucking the national trend. To try to understand the reasons for this, the Treswell group are using PIT tags to monitor visits to their two feeding stations, and they also intend to fit readers to their Marsh Tit nest boxes. Pit tags are ideal for use with the Marsh Tit. Its average lifespan is two years, so the tags should outlive the majority of the birds. The species is not common, so only a limited number of readers will be required. Even though the Marsh Tit is small, weighing in at only 11 g, the tiny bands add only an additional 0.84% to its body weight, equivalent to a human who weighs 80 kg permanently carrying a half-litre bottle of water attached to their leg.

Gadgets for administration and learning

In addition to gadgets that help citizen scientists to observe animals, plants and microbes there is another range that assist in the recording, collation, analysis and dissemination of those observations. These gadgets are without doubt responsible for the biggest boom in citizen science, especially that of the high-volume, crowdsourcing kind. They arrived when the power of the computer, the database and the network were combined into cloud computing. Networks and databases are the underlying technologies that give us the Internet, interactive websites and social media platforms, and today we can carry into the field in our back pocket a computer that is around 900 million times more powerful than the ones which guided *Apollo 11* to the moon. That computer is, of course, the smartphone. As soon as these technologies became widely available to the public, they were adopted to support citizen science.

From the perspective of citizen science, social media platforms such as Facebook, Instagram and X (formerly known as Twitter) provide basic unstructured support, but they have the great advantages of being extremely easy to set up. They also have a massive body of users who are already very familiar with the way that they work. They take care of account set-up, managing subscribers, followers and members, allowing subscribers to post images and text which will be notified to other like-minded subscribers, managing threads of comments, and even suggesting an account to potential subscribers who have shown an interest in other similar accounts. For citizen science these features allow individuals or groups to inform their subscribers of upcoming events, such as field meetings or talks, and to broadcast the results or short articles. Primarily, of course, they allow subscribers to post images of the latest thing they have seen, usually with their idea of what it is, and often asking for confirmation. Herein lie several problems, the worst one being misidentification.

Specialist accounts – for example, the UK Hoverflies Facebook group – are usually inhabited by experts, mostly amateurs, who will verify or correct an identification. However, more general accounts and personal accounts are less reliable, and even on the specialist accounts a viewer must be prepared to read through the comments to find the correct identification. Furthermore, browser searches for images of a particular species can return an incorrect image, even from specialist sites, because the browser can use the identification given with the original posting. If the social media sites are used to harvest formal records, there is a further problem. This is the tendency for users to post the big and colourful species most frequently, thus potentially distorting the overall view of diversity, and to omit the essential information on location, date and time which is needed to create formal records.

Despite these problems, which are really just challenges for the experts to overcome, the social media sites have been a great boon to natural history. They have done exactly what they were designed to do. They have generated interest and grown communities, reached out to new members and brought awareness of nature to the wider public. For instance, who would have thought that there were 1,300 people in the UK who are interested in hoverfly larvae? At the time of writing, BirdGuides has 81,000 followers on X and 21,000 followers on Facebook. The BSBI's Wild Flower Hour group has a combined following on X, Instagram and Facebook of 41,000, whereas the actual BSBI membership is less than 3,000. The Penparcau Wildlife Group account on Facebook has been important for recruiting volunteers and involving the local community. The Penparcau project started in 2015, by 2018 it had 200 followers and by early 2025 there were nearly 900. This wider interest also brings the advantage of broader geographical coverage than can be managed by the few experts, and it occasionally excites everyone with the discovery of a new taxon for the county or country. A lovely example of this occurred in 2021, when Jeff the cat caught a shrew that seemed a bit unusual to his owner, so she posted a picture of it on Facebook. This was picked up by an ecologist who thought it had the Clanger-like face of the Greater White-toothed Shrew *Crocidura russula*, a species not known from mainland Britain. Later DNA analysis confirmed this identification, which made it the first new mammal species to be found in mainland Britain for 100 years.

If they are to achieve more than the unstructured support that is offered by social media, organisations and societies have to build and maintain their own interactive websites and databases. There are a large number of these with varying degrees of support for citizen science, ranging from simple help with identification to creation, management and recording of complete projects. iSpot aims to be a one-stop shop for identification. Images of any organism can be uploaded and the iSpot community will attempt to identify it. This has the advantage that the user does not have to search for the particular social media group that deals with that organism, but the disadvantage is that the identification will be less certain, because any member of the iSpot community can make an attempt at it. Then there are the websites that support crowdsourced surveys, such as the RSPB's Big Garden Birdwatch, and Butterfly Conservation's Big Butterfly Count, in which users enter their observations directly into the organisation's database. There is no verification of identification, but crowdsourced projects are usually looking for easily identified species, and sheer weight of numbers does a lot to counter misidentification. Verification of identification is a theme running through all of these data sets. As we discussed in Chapter 3, the data must be accurate, as they are used for the scientific study of distribution, diversity and ecology, and the results are often utilised by government to guide policy decisions.

At the next level of website/database complexity is the distribution database, which is used to map the occurrence of species across the country. These databases are usually developed and run by organisations that specialise in a single type of organism, such as the BSBI for higher plants, the Dipterists Forum (DF) for true flies and the British Lichen Society (BLS) for lichens. Most of these do not allow direct input of records into their databases, but instead have a gatekeeper who verifies records before adding them to the database, to maximise the accuracy of identification. For the BSBI and the BLS the gatekeeper is the county recorder, and for the DF it is a countrywide recorder for each of the true fly families. The recorder for the Pipunculidae (commonly known as big-headed flies) does not even like to receive text records, even from expert dipterists, but prefers to be sent specimens to dissect and thus formally verify the species in this difficult family. The rewards for citizen scientists who submit records are the distribution maps, descriptions and details of species and their life histories, which are published on the websites. For example, the BSBI website has an interactive, dynamic map which takes records directly from the live database, and it also has a freely available online version of *Plant Atlas 2020*, which gives a description, a distribution map, the conservation status, a photo gallery, the frequency trend, and phenology and altitude data for every species of flowering plant found in the UK and Ireland.

Like iSpot for identification, there are websites set up to receive records for all types of organism. iRecord is a UK-based system, whereas iNaturalist was developed for reporting in the USA but is growing to encompass the world, and is currently used by 7.5 million people. The advantage of these sites for the citizen scientist is that they do not have to find the appropriate recorder for any particular organism, and all of their records are submitted in the same format with the same metadata. In addition, they can keep all of their records together online. The difficulty for specialist databases that deal with a specific type of organism is again verification of identification, and the extra work involved for the recorders for those databases. As a consequence, not all of them accept records from the generalist sites yet. In principle, in order to develop automatic feeds from the generalist sites, the specialist sites must highlight as needing further investigation all records of species that are considered difficult to identify, or which are reported from outside their expected geographical range. The appropriate recorder must then work through these flagged records to verify, correct or reject them. In the long term it is the intention that the recorder's assessment is fed back to the citizen scientist who submitted the record in the first place. One final site in this huge edifice in the UK is the National Biodiversity Network (NBN) Gateway. This database takes feeds from all of the specialist sites so that researchers have a single site dispensing accurate data for biodiversity studies. In turn, the NBN Gateway is a node in the Global Biodiversity Information Facility.

There are yet more websites that combine the data sets from national databases to give a regional or worldwide picture with some fascinating results, especially for migratory animals. The EuroBirdPortal pools data from birding websites in Europe and maps species distribution across the region in almost real time. Birders can, for instance, watch the progress of Waxwing *Bombycilla garrulus* moving down through southern Scandinavia towards the UK and central Europe as winter progresses, and so predict their appearance in the country. Big landfalls of the Waxwing can be expected if it has been a successful breeding year (indicated by the density of records), the winter is severe and the year has produced only a poor crop of fruit. It is also possible to replay a year's worth of observations to watch, say, the ebb and flow of summer migrants moving to and from southern Europe and Africa. eBird performs a similar pooling of observations for birds across the globe.

Other websites offer enhanced technical support to the citizen scientist, and these are increasingly adopting AI to do a lot of the spadework involved in examining observation data. This is particularly true for camera trap footage and audio recordings. One of the bugbears of manual examination of camera trap footage is excluding images with no animals or, conversely, spotting small animals in an image. eMammal's online analysis tool compares segments of contiguous frames looking for changes in background, which it equates to movement, and draws a bounding box around the area of motion so that a human need only focus on the box's content. eMammal's analysis tool has also learned to recognise humans and it automatically excludes images containing them. Similarly, Wildlife Insights has built an AI system and trained it to remove all images that it is confident contain no animals, and for those images that do contain animals to suggest an identification at the lowest taxonomic level of which it is confident. Thus, for example, it might just suggest deer rather than Red Deer if it is certain that the animal is a deer but it cannot recognise the species. Clever conventional programming can also help to improve the accuracy of identification from images examined by citizen scientists by arranging to present every image to multiple people. If a set majority agree on the identification it can be accepted, otherwise the image is passed to an expert for examination. The BTO's Acoustic Pipeline, which we discussed earlier, is an online application that analyses continuous full-spectrum recordings of inaudible, ultrasonic animal calls, predominantly those of bats. It can distinguish multiple species calling simultaneously, separate bat calls from those of other animals, and classify the calls into echolocation, social contact or feeding buzzes. It also gives a warning if it is a rare or unexpected species, and when its confidence in its identification is low.

For citizen science, the most sophisticated websites are those that run complete projects, allowing the project initiator, who is not necessarily a professional scientist, to create the project, attract and train their citizen science colleagues and organise data collection by them, supporting data input and analysis and sharing the results. eMammal is one such website, and its UK-originated equivalent Mammalweb is another, but the biggest and most prominent of all is surely Zooniverse. This amazing site, which only started in 2007, had signed up in excess of 1.6 million citizen scientists by 2019. It started, as its name hints, with astronomy. An Oxford University research student was faced with having to scan vast numbers of images looking for galaxies, with the aim of classifying them into different types. He managed 50,000 in a month, working 12-hour days, and it was clear that this was an impossible task for a lone individual. Thus the first project, Galaxy Zoo, was created. At the end of the first year 150,000 volunteers had classified over 50 million galaxies. More astronomy projects followed, and the platform now covers multiple disciplines, including history, the arts, social sciences and, importantly for us, the life sciences. So if you get bored classifying galaxies, you could spend your time transcribing extracts from medieval records or classifying dolphin vocalisations. At the time of writing, Zooniverse has 90 live projects, more than 50% of which are life science based, and it has set up almost 500 projects since its inception. It works hard to maintain the engagement of its citizen scientists. Researchers commit to being active on a project's discussion board and regularly posting blogs. Many participants have been co-authors of publications about projects where they have been involved in analysis and discussion.

FIGURE 5.5 A few of the many nature projects on offer on the Zooniverse website. (www.Zooniverse.org)

Volunteers are also sometimes able to participate in spin-off scientific tasks, and are actively encouraged to provide feedback to the development team.

The most ubiquitous of the modern gadgets for citizen scientists is the smartphone. Almost everyone owns one. It combines a powerful computer with a large amount of storage, ready access to all of the resources of the citizen science websites from almost anywhere, a pretty good camera, accurate GPS and a long battery life, and yet it is smaller and lighter than the average field guide. All one has to do is download the appropriate apps, and there are lots of them. One common category of app is aimed at replacing the field guide, or at least pointing the field naturalist to the right section of a guide. Some of these identification apps, such as iNaturalist and Seek, are generalist and cover all types of organism, whereas others focus on particular types, such as Flora Incognita for flowering plants and Merlin Bird ID for birds. They are mostly powered by AI, which learns from the huge volumes of digital images or audio files, for those like Merlin Bird ID (which can recognise song) that now exist in the cloud. Their veracity depends on the accuracy of identification of the media from which they are learning, and the quality of the digital information that they are asked to identify. Clearly those powered by AI that has learned from libraries curated by experts will be more accurate. For instance, Flora Incognita's algorithms were trained, with guidance from botanical experts, from a library of over one million images. This particular app also uses location, from which it can look up soil and habitat type, and time of year, from which it can look up a species' seasonal presence when coming to a decision.

After a shaky start, these apps have improved enormously and for many species they are very precise – Flora Incognita tested against untrained images reviewed by experts currently has an accuracy of more than 83%. However, they will never be perfect, partly because they are dependent on the quality of the image they are being asked to identify, in terms of clarity, lighting and shooting angle, and also because many taxa require dissection, high magnification or exact measurement for a definitive identification. Many insect species also present these problems with precise identification, and as they also tend to be fast moving, image quality can be a problem, too. In fairness, the apps usually give the user a percentage confidence, a ranked list of possibilities or an answer at the lowest taxonomic level at which the app is confident. One thing that they are not good at is helping people to distinguish between species for themselves. For this a key or field guide is an option, but it will always be the case that the most effective and enjoyable way to learn identification is to go out into the field with an expert. When choosing an identification app, it is best to use one of those recommended by the society that specialises in the type of organism you are interested in, as they will have made a skilled assessment.

There are apps available for sound recording, too, so that this can be done in the field. iRecord and iNaturalist have them as well as their websites for the input of records. Flora Incognita automatically stores all identifications that its users confirm. Some apps that were not designed specifically for citizen scientists are useful for more traditional recording, where the data are destined for specialist databases and cannot be entered directly. The most important of these are the ones that provide

FIGURE 5.6 Just two of the many smartphone apps designed to help and support the contemporary citizen scientist. The Flower-Insect Timed Count (FIT Count), developed by the UK Pollinator Monitoring Scheme (PoMS), is used to record timed flower–insect interactions, whereas Flora Incognita uses AI to aid identification of plant species from photographs taken by the citizen scientist. (UK Pollinator Monitoring Scheme, 2024, Flora Incognita project)

an accurate map reference, such as Grid Ref OS and OS Locate. Finally, there is a plethora of apps for citizen science projects with a more limited scope. These still mostly fall into the category of recording apps, and they are often related to environmental damage. For example, in the UK, Ozone Injury is used to report indications of ozone injury to plants, and Rare Arable Flowers helps people to identify and report on the location and abundance of this declining group of plants. In the USA, Project Budburst records phenological events. At a global level, Secchi is a system for mariners to report on the abundance of phytoplankton, and Natura Alert is used to report on habitat damage or other threats to nature.

Futures

Citizen science has already reached into the future with gadgets that employ AI for such things as auto-identification and filtering camera trap footage, and close behind there is another futuristic technology that it is beginning to explore. This technology has been the province of professional, laboratory-based scientists for some years, but as research has progressed, costs have dropped and techniques have improved to bring DNA sequencing within reach of the serious amateur, at least in its simpler form of DNA barcoding. Fragments of DNA consisting of around 600 base pairs, if taken from a gene with minimum intraspecific variation, can be used to identify, or barcode, individual species. Suitable genes have been adopted for the various kingdoms of organisms. For animals, a portion of the cytochrome C oxidase I gene from mitochondrial DNA is used, whereas in fungi the Internal Transcribed Spacer gene in ribosomal RNA is most commonly used. The search for a suitable barcode in plants took longer, but seems to have settled on portions from a pair of genes. Libraries of these barcodes or even complete genomes are being built up. For example, International Barcode of Life (IBOL) is specifically for barcodes, the Darwin Tree of Life is working towards a library of full genome sequences for 70,000 organisms in the UK, and GenBank is an international library of all publicly available DNA sequences.

Only a small amount of material is required for DNA analysis. Its DNA is extracted, and the target sequences are then bulked up by the polymerase chain reaction, checked for purity, sent for sequencing and finally compared with library material. Citizen scientists can equip themselves to do everything except the sequencing. At the time of writing a commercial kit costs around £2,000, much the same as the cost of a decent microscope or camera, and the costs per sample currently run at around £2 for reagents and £5–10 for sequencing. The expectation is that the costs of sequencing could fall to as little as 10p per sample as demand grows. The good news is that there is no need for a laboratory – everything can be done in the citizen scientist's kitchen. DNA barcoding is definitely not a project for high-volume citizen science. A considerable degree of training and practice is required before extracts of high enough quality for sequencing can be prepared. This might take anything from two weeks for very adept individuals to six months.

This is all well and good, but what does DNA barcoding give the citizen scientist? Apart from the fascination and enjoyment that the subject will engender in some, the technique will be revolutionary for the citizen scientist who is struggling to disentangle cryptic, morphologically similar species groups. This is particularly true for mycologists and lichenologists. Only 5% of the estimated three million species of fungi have been described, and many of those are exceedingly difficult to identify based on morphological characteristics alone. Amateurs in the USA have led the way with mycological barcoding, but in recent years citizen scientists in the UK have also started to adopt this approach. For instance, during the COVID-19 lockdown, members of the Sussex Fungus Group collected spring-fruiting fungi from their local area and subjected problematic examples to DNA barcoding. They found several species new to Britain and also a previously unknown link between the sexual and

FIGURE 5.7 DNA barcoding in a home laboratory. The Hampshire Fungus Recording Group using a Bento Lab to sample and extract DNA from specimens after training by the Lost and Found Fungi citizen science project. (Brian Douglas)

asexual stages of a species. In other studies they have found more species new to Britain, some new to Europe and potentially to science, and they have published half a dozen papers describing their finds.

DNA barcoding can also be used for ecological studies – such as for determining the identity of pollen that is being carried by insects, or which insects have visited flowers (they leave traces of DNA behind) or what foods animals have been eating (from DNA in their stomach contents or faeces). When organisms are identified from a sample containing DNA from several species this is called metabarcoding. It is metabarcoding that is used for environmental DNA (eDNA) studies, in which environmental samples collected from a particular medium, very often water from rivers, ponds or the sea, are analysed by barcoding to discover what organisms are present. The techniques involved are somewhat more complex and are still laboratory based. However, there have been projects in which citizen scientists simply helped to collect the samples, which in itself is an exacting process. A study of the monitoring of the Great Crested Newt *Triturus cristatus* in ponds around the UK tested both the effectiveness of eDNA as a marker of their presence and the effectiveness of the citizen scientist in collecting suitable water samples. Both were found to be very effective, and this method markedly reduced the effort required compared with traditional techniques such as bottle traps and torch searches. However, the eDNA was not so good at estimating abundance.

Chapter 6

Becoming a Citizen Scientist

When social scientists have investigated citizen science projects, they tell us that a citizen's motivations for getting involved with science are wide ranging. They may take it up for altruistic reasons, out of concern for nature, to share their knowledge or to encourage an appreciation of nature in their children. They may take it up for social reasons, to meet likeminded people or for more self-centred reasons, perhaps hoping to benefit their future careers. Very many take it up simply because

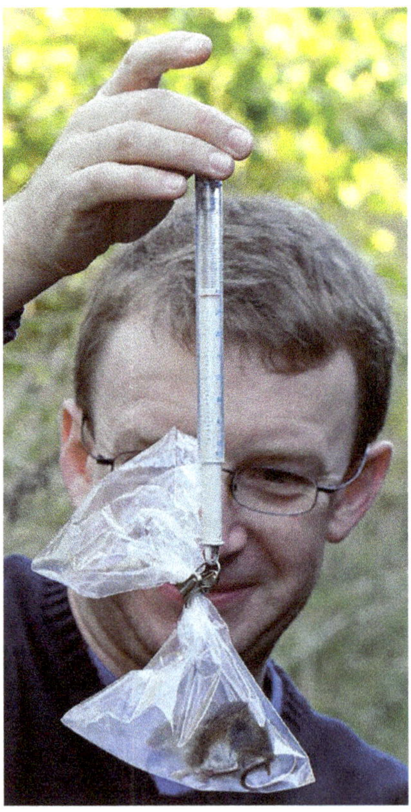

FIGURE 6.1 Citizen scientists in action: weighing a Wood Mouse *Apodemus sylvaticus* during a small mammal citizen science project. (UKCEH)

they find it interesting or satisfying. Of course, everybody is different. Most people will have a number of different motivations and these can change with age – the young are more likely to be looking to boost their career prospects, whereas older people are more likely to be motivated by the desire to share their knowledge. The benefits for the citizen scientist are equally wide ranging, and most are realised when the citizen scientist's original goal has been reached, or it may be an ongoing process of continuing input and benefit. The benefits for the professional scientist who is organising projects are fairly simple and straightforward. At its most basic, citizen science is an economic method of large-scale data collection that would not otherwise be possible. It has taken some time for professional scientists to realise that for project continuity they must ensure that their citizen scientists also derive some benefit from the project.

Two very important outcomes for the citizen who is getting involved with the life sciences are an increase in wellbeing, as there is an increasing acceptance that getting outside and communing with nature are beneficial for mental and physical health, and the unlocking of the mystery of the practice of science. Too often science is presented as a closed shop available only to the guild of highly trained professional scientists with their lab coats, safety specs and access to expensive laboratory equipment, and as being beyond the ken of the ordinary citizen. Becoming familiar with the way that the scientific mind works will help enormously with interpreting what professional scientists tell us, or more usually what the press tells us the scientists have told us. This will help to dispel the two extreme attitudes to science and scientists – that everything the professional scientists say is fact, or that professional scientists are only presenting the data that they want you to see.

Thus the benefits of involvement are considerable, and having read this far and seen the many significant contributions that citizen scientists have made to the life sciences, we dare to hope that you have been inspired and are now asking yourself whether you, too, could do some citizen science. The answer is, of course, yes, anyone can, even if it is just adding a small piece of data to a large national or international survey. How to do citizen science is the subject of this chapter.

You might also ask if there is anything left to do that does not require laboratories and expensive equipment. James Lovelock was adamant that 'there is still a place for the amateur scientist', citing the lone observer of the desert night sky who was the first to witness the approach of comet Hale-Bopp in 1996, ahead of the professional observatories. There are many more examples, such as Steven Vogel, a biologist with expertise in insect aerodynamics, who studied and published on the subject (rather overlooked by professionals in the fields of botany and forestry) of mechanical adaptations of trees to resist storm damage, and David Wilkinson, a science professor whose areas of study include testate amoebae. Both of these scientists have written that there is great scope for amateurs working avocationally to add to our knowledge of these subjects. This is especially true for ecology, where diversity is so great, even in the UK, that keen and close observation can answer many questions. A recent example of this would be the nesting behaviour of Black Seabream *Spondyliosoma cantharus*, which was observed for the first time in images taken by citizen scientist divers. The fish clears stones and other loose debris so that the female can lay her eggs on the clear bedrock of the 'nest'. Add to this the fact that nature never stands still. Change is afoot with the warming climate, and citizen scientists are as likely as professionals to observe early indications of it by reporting such rare events as the arrival of new species in their local area.

Passive or active participation

You can easily participate without having to create your own project. If you follow this route you will most probably be joining in with one of the mass-participation, crowdsourced projects. Doing so will save you having to do all of the design, organisation and other tasks required to create your project, which we shall discuss later, but it does not mean that you will be restricted to simple, short, observational studies such as happens in the Big Butterfly Count or the Big Garden Birdwatch. The large, nationwide projects offer a whole range of different levels of involvement, even the possibility of performing your own experiments.

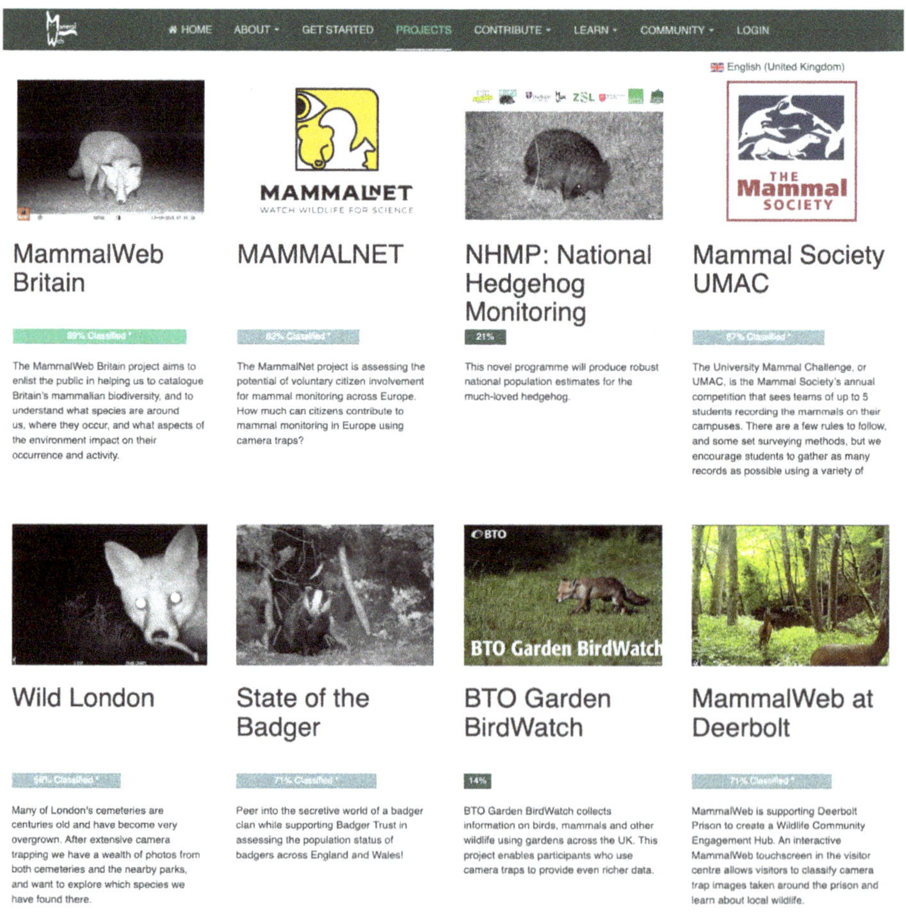

FIGURE 6.2 The projects page of the MammalWeb website, showing the projects available for citizen scientists to join. (www.mammalweb.org)

The Conker Tree Science project was set up specifically to go beyond using citizens in long-term, large-scale studies of species distribution or abundance. Rather it sought to involve them in answering ecological questions or, as the professional scientists put it, in hypothesis-led citizen science. The Horse Chestnut Leaf Miner *Cameraria ohridella* is a moth whose larvae mine the leaves of the Horse Chestnut, or Conker, tree *Aesculus hippocastanum*. It has spread rapidly across Europe from the Balkans, arriving in the UK in 2002, and has become a pest. Heavy infestations can turn most of a tree's leaves brown by mid-summer. When the project started, in 2010, the miner was relatively new to the UK, still at the margin of its distribution, and the project was interested in the dynamics of its spread and the response of its predators, which are parasitoid wasps. Two hypotheses were investigated – first, that the amount of leaf damage and secondly that the amount of parasitism of the miner by wasp larvae would be largest in places where the miner had been present for longest. To investigate these hypotheses the citizen scientists were asked to assess leaf damage against a pictorial score card during the summer (mid-June to the end of September) and, more interestingly from the perspective of this chapter, to raise and report on the occupants that emerged from leaf samples taken in the first week of July. To do this the sample leaves were stored in sealed plastic bags for two weeks and the emergent insects were placed in one of three categories – the leaf-miner moth, a parasitoid wasp or other.

Using mathematical modelling, based on previous records of occurrence of the leaf miner in the UK, the probabilistic rate of short-distance dispersal found from its recorded spread across Europe and feedback from the data itself to account for the rarer occurrence of long-distance dispersal, the scientists analysing the data did find that damage was worse where the leaf miner had been present for longer. They also found that damage increased rapidly for the first three or four years and then levelled off, and that it was most likely that there were two generations per year rather than the three generations found in parts of Europe. Over the years the amount of parasitism increased from 1.6% to 5.9%, so again it was highest where the leaf miner host had been present for longest, and it was attributed to native generalist parasitoid wasps. A study in Europe that took place over 20 years found that increasing predation was by more specialised, host-tracking parasitoids, and this led to the interesting speculation that predation occurs in two phases – first by native generalist parasitoids, and then over the longer term the adaptation or tracking of specialist species.

More than 3,500 citizen scientists took part in the Conker Tree Science project in 2010, a large proportion of whom were schoolchildren. In fact it was schoolchildren who returned more than 80% of the records on reared insects. In this respect the project is reminiscent of the more recent hypothesis-led citizen science project that we described in Chapter 3, where Swedish schoolchildren studied the accumulation in fungi of radioactivity from the Chernobyl fallout. The popularity of these two projects suggests that many citizens are keen to be involved in these more experimental studies.

Another such project which was quickly oversubscribed involved growing Broad Bean *Vicia faba* plants and manipulating their pollination. The scientists from the

University of Sussex who set up the Bees 'n' Beans project were actually interested in seeing if the methodology was suitable for directly monitoring pollinator deficit around the country. Pollination in crop plants very often requires animals or is improved by them, and in temperate climates those animals are mostly insects, especially bees and flies. In Europe, 84% of crops species are dependent to some extent on insect pollination. Yet, sadly, the latest estimates indicate that insect numbers have declined dramatically in recent years. Although many studies use a count of pollinators as an index of pollination potential, very few directly measure the level of pollination. In the Bees 'n' Beans project, participants were given a project kit consisting of seeds, pots, exclusion netting, instructions and recording sheets. They had to grow the beans and then, just before flowering started, select four plants of equal stature for the experiment. One of them was covered by the exclusion netting to keep out the local pollinators, the second plant was manually pollinated using pollen from the fourth plant and was also available to the local pollinators, and the third plant was left solely for the local pollinators. All of the participants lived in an urban setting and ran the experiment there, usually in their own garden.

Although the professional scientists were interested in the effectiveness of the methodology, the hypothesis that the experiment was designed to test was that there were insufficient local pollinators to maximise pollination in broad beans in the urban environment. For the null hypothesis to be true, the plants covered by netting would have to produce fewer beans than the other two plants, and the plant left solely for the local pollinators would have to produce fewer beans than the manually pollinated plants. The covered plants did indeed produce fewer beans, but there was no difference between the manually pollinated and insect-pollinated plants. The results indicate that broad beans do require pollinators to maximise crop yield but that, in the year of the experiment and in urban spaces, pollinators were not limiting. Initially, 551 volunteers signed up. Of these, 176 remained engaged until the end, although only 80 (14.5%) successfully completed the experiment. The failures are equally important because their feedback helps to improve the experimental design. In fact the researchers at the University of Sussex were sufficiently confident in it to put this methodology forward as suitable for a national programme for monitoring pollinator deficit.

Britain and Ireland have the most detailed database of plant distribution in the world thanks to the thousands of volunteers who record the species that they see on walks and field trips. This sort of opportunistic collection of species occurrence data introduces biases which make it difficult to monitor species decline or spread except in terms of broad trends. Also it cannot tell us anything about species dynamics within individual habitat types – that is, whether certain habitats are faring better or worse. To overcome this gap in our knowledge, a more systematic approach that monitors abundance on a small scale in fixed locations and various habitats over many years is needed.

The National Plant Monitoring Scheme (NPMS) was set up by BSBI, the UK Centre for Ecology and Hydrology, Plantlife and the Joint Nature Conservation Committee to do just this. It is aimed explicitly at citizen scientists who want to understand more about the science of ecology and to gain a more intimate knowledge of our semi-natural habitats. It was recognised that for the project to work well, it would be best if the citizen scientists remained engaged for many years, and to this end citizen scientists were recruited to participate in the early trials of the experimental design and to refine it over time. In essence, the project works by allocating the volunteers a monad (a 1 km square) close by, within which they select up to five small plots in a range of habitats. They visit these twice a year to record the abundance of the plant species growing there. In choosing the plot locations, deciding on the plot layout and seeking any permissions required (if the plot is on private land), participants tend to take personal 'ownership' of their plots. The biannual detailed abundance survey gives them an intimate knowledge of the dynamics of their small plots, as well as contributing to the countrywide view. The NPMS was started in 2015, and by 2021 as many as 1,700 volunteers had been allocated monads and data had been returned for almost 1,000 of these, encompassing 4,500 plots. As the scheme continues there are plenty of monads left to be allocated – all you need to do is sign up.

Our last example of passive citizen science projects is particularly beneficial for people who find themselves stuck indoors for whatever reason. The Open University offers a wide range of free virtual experiments which can be done simply out of interest or even be used for self-training. For instance, someone venturing into the virtual laboratory can work out the quantity of cattle and pasture needed to sustain a bat colony, another virtual experiment involves quantifying the amount of DNA in a sample and quality checking it using UV spectrometry, or you could determine the evolutionary relationship between a collection of hominid skulls.

In reality it is not possible for an individual citizen or even a local group to set up such big nationwide projects. They tend to be run by professional scientists backed by large institutions, charities or societies, and for very good reasons. The experimental design and data collection methodology need very careful consideration and ample experience of scientific practice. This is to ensure that the results will adequately test the hypothesis, especially if the project involves many contributors over a long period of time. For example, during the development of the NPMS the professionals chose the pool of monads using fairly complicated maths to introduce a habitat bias into the score given to every monad in the country, so that monads with larger areas of rarer habitats were more likely to be chosen from a subsequent random weighted draw. Habitat types had to be carefully chosen and filtered to produce a list of semi-natural habitats that could be readily identified on the ground by the volunteer surveyors. Highly modified habitats (such as brownfield sites or improved grasslands), mobile or shifting habitats (such as mud flats or strandlines) and dangerous habitats (such as cliffs) were excluded. Some habitats were aggregated – for instance lowland beech and yew woodland, lowland mixed deciduous woodland, upland oakwood, upland birchwood, and upland mixed ashwood into Dry Deciduous Woodland. Next, suitable indicator species had to be chosen. These needed to include commoner species that could be identified by less experienced surveyors. Positive indicator species for each habitat were chosen using another biased weighted random selection technique from the pool of species listed in a habitat's constituent National Vegetation Classification (NVC) communities. These were not the only design choices, as consideration also had to be given to plot size and shape, plot placement, minimising plot selection bias and so on. All of these points show just how much thought needs to be given to the design of these big national projects.

Such projects are also costly and time consuming, although from the perspective of the professional they are both economic and time efficient! A study in 2012 of a wide range of citizen science projects found that the annual cost of the type that contribute to policy development, usually nationwide surveys of some aspect of biodiversity, was around £100,000. This is actually a small fraction of the worth of the in-kind contribution of the volunteers doing the surveys, hence their high value compared with cost. The costs of the Bees 'n' Beans and Conker Tree Science projects were more modest but still significant. The hard costs – that is, costs excluding the professionals' time – for the Bees 'n' Beans project were around £2,500, mostly for printing, postage, netting and the website, and for the Conker Tree Science project were considered to be equivalent to the cost of a postgraduate student working for 18 weeks.

Nationwide projects can evolve out of studies started by individuals, but more by accident than design, in the sense that the originators had not envisaged the scale of the project. This is probably true of many of the long-standing distribution surveys. The Beach Patrol Scheme started by the Ornithological Society for New Zealand (OSNZ, now called Birds New Zealand) in 1951 is a good example. New Zealand is surrounded by thousands of kilometres of ocean, and the closest continental landmass is Australia, 1,600 miles away. Sea birds are great wanderers, and many species from all over the southern oceans visit the region, but very often their presence is only known from the corpses found on New Zealand's beaches, not least because many species are extremely difficult to identify when free flying at sea. Before the Beach Patrol Scheme started, a few OSNZ members would sometimes opportunistically walk stretches of beach, recording the corpses that they found and reporting them in *Notornis*, the society's journal. There are four problems with these data – only a tiny proportion of the coastline is covered, there is a tendency only to report the rarities, sometimes no rationale is given to support the identification, and the length of beach covered is not usually stated. Nonetheless, the data give some broad indication of the seabird distribution before 1950. They also show the value of the personal observations of citizen scientists, so long as these are published and not allowed to languish and be lost.

The Beach Patrol Scheme stuttered somewhat after its first introduction and was relaunched in 1960, since when it has gone from strength to strength. In the late 1990s nearly 4,000 km of coastline were being patrolled every year, producing reports of 5,000–14,000 dead birds, and by 2022 the database contained 30,000 records of around 440,000 corpses, compared with just under 7,000 specimens found between 1939 and 1959. All of these data have been put to good use, being included in several published papers – for instance, as evidence to support the rejection of abnormal weather events as a possible cause of the changing migration pattern of the Thin-billed Prion *Pachyptila belcheri*, and as confirming the serious decline of the Sooty Shearwater *Ardenna griseus*.

Although it is not practicable for the lone citizen scientist or local groups to embark on nationwide projects, there is nothing to stop them setting up their own local project within a national project. This sort of half-way house between passively joining one of the big science projects and actively creating your own has several advantages. The data to be collected and the methods of collecting them will already have been carefully considered and set out, and training, data collection and collation support, such as a ready-built website, may be available, too. All that may be required is a few tweaks to the national protocol to suit the particular project you are envisaging. If you are using the big project's database, be sure to check beforehand that you will be able to access your data for analysis. Not all national projects have this facility.

NPMS+ is an excellent example, as it was created specifically to explore how local groups could develop their own projects using the NPMS resources, particularly the well-developed and proven protocols for monitoring habitats and biodiversity. Surveyors had expressed interest in using the methodology for locations outside the NPMS pool of monads. In the pilot, local groups are allowed to select the monads of interest to them and to add habitat and species options to suit their project's needs. The Plant Portal website facilitates administration of the project, such as setting up squares, managing participants and accessing data, and it enables volunteers to select their monads, choose plots and enter data. In addition, all of the training material created for NPMS is available, as well as new material specific to NPMS+.

Another example would be pollinator monitoring in your garden, local park or, as a small group, for your local nature reserve. There is a nationwide scheme for this, the UK Pollinator Monitoring Scheme (PoMS), but there is nothing to stop an individual or group doing their own study while contributing to the nationwide effort. Although the methodology is deceptively simple – a participant counts the number of pollinators landing on flowers in a small area over a ten-minute period – a great deal of thought has been put into the method and the data to be collected, as we shall see later. The close observation required for PoMS could well lead the enquiring mind into a more experimental approach, such as the broad bean pollination study we discussed earlier.

It was detailed observation by means of close-up photography that led one of us to become involved in a series of experiments on the pollination of the native orchid, Broad-leaved Helleborine *Epipactis helleborine*. One of the photographs captured a social wasp, its head covered with pollinia (clusters of the orchid's pollen grains), visiting one of the helleborine's flowers. Reading around this, it transpired that although social wasps are the primary pollinators of Broad-leaved Helleborine, in some parts of the world this species only self-pollinates, in some it must be insect-pollinated and elsewhere it can be insect- or self-pollinated. I wanted to know what mechanism this plant uses in the UK. The outcome was that, in my garden at least, Broad-leaved Helleborine does not self-pollinate and needs the wasps to set seed.

Once you start trying to interpret your own observations you are turning to active citizen science. Now you need to start thinking in a scientific way, posing questions and collecting information to answer them. In short you need to adopt scientific methodology. This has evolved over the centuries, but basically involves making your own detailed, contemporaneous observations and applying a healthy degree of scepticism to your interpretation of the results, so that the conclusions from the

FIGURE 6.3 The Broad-leaved Helleborine *Epipactis helleborine* experiment in progress. The plant in the foreground is covered with a muslin bag to exclude pollinators, while the uncovered plant in the background is being visited by a Common Wasp *Vespula vulgaris* with the orchid's pollinia attached to its head.

work are not biased by any of your initial assumptions. The general approach should be to make observations and read around the subject, to come up with a question or putative interpretation and from this to create a hypothesis that can be tested to confirm or refute the original conjecture. A very elegant example of this process is provided by the experiments of Darwin and his son on phototropism. They observed that when a young grass shoot (known as a coleoptile) is illuminated from one side it bends towards the light. The bend does not occur at the tip, but lower down. The question they asked is what part of the plant controls the ability of the coleoptile to bend. They did several experiments. When they either cut off the tip or covered the tip with foil the coleoptile did not bend. However, when they covered the lower part of the coleoptile with foil or the tip with transparent material, the coleoptile did bend. Darwin was able to conclude quite convincingly that 'some influence is transmitted from the tip' and that this caused the coleoptile to bend.

Hypothesis and design

Throughout this and previous chapters we have referred to the idea of a testable hypothesis, so before we move on to data and the control of data errors it is worth explaining this concept. Usually, in order to prove an idea to be correct we might think that all we need is corroborating data, but this is not so. To be certain of what our data have shown we need to set up a null hypothesis that can be disproved.

The logic for this is deceptively simple. You cannot prove a negative, but it can be disproved. So if you start off by saying that there is no pollution in a river and then find some, the null hypothesis is rejected and you can say the river is polluted. If you start by saying the river is polluted but do not find any pollution, this does not tell you whether the river is or is not polluted, it may not be or you may not have looked hard enough. The details and nuances of the null hypothesis go deeply into statistical processes that are necessary when large amounts of data are being analysed, but for the majority of its use in citizen science projects this is enough to help in appreciating errors that may crop up from time to time.

In terms of scientific problems, data errors can be devastating if they are not controlled. They fall into two broad groups, which are called type I errors and type II errors. Type I errors, also called false-positive errors, occur when the null hypothesis is true (e.g. there are no crocodiles in the pond) but is erroneously rejected. This is an extreme example, but assumptions can be made that create a preconceived idea which drives the study forward to a specific outcome, usually one that is desired for reasons beyond simple investigation. Type II errors, also called false-negative errors, can occur when the null hypothesis is true (e.g. the ground does not get wet when it rains). Again this is an extreme example, but these types of error can result in an assumed causal relationship when in fact one does not exist. This kind of error is fortunately rare in citizen science projects because there are usually a number of people all scrutinising the same data, which makes misinterpretation both difficult and unlikely.

Creating a testable hypothesis is not necessarily easy, but it does fit in with another aspect of citizen science which you as an individual may wish to encompass. Quite simply, do you have a wider plan? You may have started out with a particular goal in mind, but investigating that could well lead on to other areas and investigations. Try to plan ahead as much as possible for just such redirections, and if they arise be prepared to sit down and reformulate your plan. Serendipity can be very productive, but if you can help it along with a flexible attitude, so much the better.

With your testable hypothesis sorted out you can set about designing your experiment and determining what data to collect. One of the basic principles of good experimental design is that there should be controls. This is the term that scientists use for experimental objects that are given exactly the same conditions as the test objects, except that they do not receive the test treatment. The results from the controls form the benchmark against which the test objects' results are compared. Without controls it is not possible to adequately evaluate the null hypothesis. This is a chasm into which politicians regularly plummet, claiming that things are better than they would have been if they had not done something or, equally, that they are worse than they would have been because the opposition did something. Since there is not a parallel universe where everything is the same except that the 'thing' has not been done, neither conclusion can be drawn. What politicians try to do to get around this, and scientists have to do in some situations, too, is to pick an analogous item that has not received the treatment to act as the control. Bias can creep in here, which scientists must overcome by setting out the precise criteria that they used to make the choice. Politicians often seem to shamelessly embrace the bias by picking a 'control' country that best suits their argument.

In Darwin's phototropic experiment the control was simply a plant grown alongside the treated plants but given no treatment. As a result, in each experimental run he could measure the degree of curvature that would have occurred naturally in the treated plants if they had not been treated. It was essential that these control plants were grown contemporaneously with the treated plants, to avoid the possibility of some other condition altering or affecting the degree of curvature if they were grown at different times or places. Similarly, if you were using the PoMS methodology to assess the effectiveness of a management treatment in increasing pollinator abundance, simply comparing numbers before and after the management treatment would not be sufficient. An increase in pollinator abundance could be due to good weather and nothing to do with the treatment. It will be necessary to leave some of the reserve untreated as a control and do a 'before and after' assessment of the untreated part of the reserve as well. If the reserve is too small to split, or it would not be practicable to split it, an area of land outside the research area must be chosen to be the control. This is where the politician's bias can creep in, so to avoid this the outside area must be carefully chosen to ensure that it is strictly analogous.

A second basic principle of experimental design is that the treatment is replicated either in space or in in time. There is no way of knowing whether an effect that is observed in a single instance of a treatment is just a coincidence, even if there is a control. If Darwin had grown just one coleoptile in each treatment, and reported his results based on this, the scientific community would not have taken his findings seriously. In laboratory-based experiments like those of Darwin, replication is usually done in relation to time – that is, the experimental results need to be shown to be repeatable – whereas in the field replication is more often done in space, with each treatment and the control having multiple plots. In our hypothetical case measuring pollinator abundance on a nature reserve, replication would involve monitoring multiple small squares in the managed and control areas. The number of plots, squares, plants – whatever the unit of treatment is – you need will depend on how variable the results are between units being given the same treatment. There is also the question of how big this variation is compared with the difference between the treatment and control. If the variation is much larger than the difference between treatments, a larger number of samples will be required for a convincing evaluation of the hypothesis. For example, if a grain of wheat can vary in weight by 40 mg and a fertiliser treatment only increases average weight by 10% (i.e. 4 mg), you will need to weigh more than one or two grains from the control and treatment plots to prove it. Not doing so puts you at serious risk of drawing a false conclusion. It may be a false-positive or a false-negative error, depending on the average weights of the few grains you weigh from each treatment.

In all experiments that are not being run in a constant uniform environment – basically all field experiments – it is vital to position the replicate experimental units at random across the experimental area to avoid expectation bias creeping in. An extreme example of such bias from our hypothetical pollinator experiment would be picking the most floriferous plots in the managed area and the least floriferous ones in the control area.

When designing your citizen science project, a primary consideration should always be how to set a standard for data being collected. Although it is possible to alleviate and on occasion eradicate statistical errors, this does make data handling very complicated. In the case of identification data, for example, without systems in place it is virtually impossible not to end up with some misleading data. This is a very serious point, as citizen science is pivotal in creating distribution and population maps and species density data. Selection of the data to be collected and the method of data collection requires careful consideration in relation to the hypothesis you are testing. Even for apparently very simple projects it is important to make sure that the data you are collecting are what is necessary to answer the questions you are asking.

For instance, the Mammal Society and other organisations have been surveying the Harvest Mouse *Micromys minutus* since the 1970s. This tiny mammal, which weighs in at only about 6 g, is rather elusive, which makes visual surveys or the use of gadgets (e.g. camera traps, humane traps, hair tubes, footprint tunnels) inconsistent or impractical. The best survey method is to search for the abandoned nests of this species in the autumn, when much of the vegetation has died down. The Harvest Mouse lives an aerial existence, and builds nests of woven grass part of the way up grass stems. In the autumn these little grass balls, which are perhaps 10 cm in diameter, can be picked out by the keen eye. Until the 2020s, records from these nest surveys simply kept presence data, merely recording that a nest or nests had been found at a particular location. When these records were used subsequently to analyse the distribution and abundance of the Harvest Mouse, the simple presence data were found wanting. Only an unreliable approximation of abundance was possible by using the ratio of harvest mice to wood mice, and, when a location was empty, there was no way of knowing whether the Harvest Mouse was actually absent from the area, the area had not been searched or, because the more you look the more you find, the area had not been searched enough. These issues have been overcome in more recent surveys by a few simple but rather clever changes to the data that are recorded during a survey. First, every survey is recorded, whether or not nests are found. Then, for every survey the number of surveyors, the time spent searching and the number of nests found are noted, and, finally, to make the data even more useful, the habitat at the site is also noted. With the new data, the rate at which nests are found can be calculated, which allows a much better estimate of abundance to be calculated and, better still, it factors out the bias that could be introduced by variation in surveyor effort between sites. The rate also allows direct comparison between habitats and thus indicates the Harvest Mouse's preferences. Collection of negative results means that areas that have not been surveyed can be identified. The rate of finding nests is also a great tribute to the dedication of the volunteer surveyors. On average a surveyor only finds one nest every 25 minutes.

Even the simplest techniques will benefit from careful advance preparation, and even if you are the sole experimenter it is worth committing your methods to paper, so that they can be replicated with confidence. Often the first attempt at something will fail because of some problems with the method. In fact this is quite a normal occurrence in science, and is a positive outcome because you can learn from the experience and tweak your technique. It would have taken Darwin some time to figure out the treatments for his phototropism experiment, even though they are deceptively simple. Finding the most sensitive age of the coleoptile, suitable opaque and transparent covers and the best light source would have involved some trial and error. Remember back to science experiments at school – you can probably count the number that worked first time on the fingers of one hand.

The Flower-Insect Timed Count (FIT Count) technique used by citizen science participants in the PoMS project is an excellent example of the care and attention to detail necessary to establish even a relatively simple procedure. In an FIT count the number of insects that land on a flower in a 50 cm × 50 cm square over a ten-minute period is counted. It takes many pages just to describe this deceptively simple procedure. The time of year and weather conditions when counting is allowed need to be defined, as do the 14 target flowers, with flowering periods spanning the allowed time of year, some of them being individual species and others being groups (e.g. buttercups). Then there is the question of how to make a quadrat (a frame isolating a standard unit of area) and how to orientate it with respect to the flowers – an important consideration as some of the flowers are trees or grow vertically, such as Hawthorn and Ivy. Participants are asked to count the number of flowers within their quadrat, and this requires the definition of what constitutes a flower when considering species with spikes, umbels and composite flower heads. The visiting insects are also divided into groups, such as bumblebees and hoverflies, but there are also catch-alls so that anything smaller than 3 mm goes into one group and anything that is bigger than 3 mm but cannot be identified goes into another. Finally, the participant needs to know what counts as an insect visit – for instance whether to count insects that are already on flowers when the count time starts, whether to count the same insect when it visits more than one flower in the quadrat, and so on.

No matter how much thought goes into the experimental design, some projects may fail because of unforeseen factors, but do not be disheartened – you will almost always learn something from the experience. This happened with another experiment involving Broad-leaved Helleborine in my garden. Following on from the pollination work, I decided to consider germination. Orchids have an unusual life history, one aspect of which is that their germination and initial growth are entirely dependent on a fungal partner. The orchid obtains nutrients from the fungus until it has grown large enough to produce a green shoot, which can take several years. In the case of the Broad-leaved Helleborine some of the older estimates suggest it can take as long as eight years, but this seems excessive, as many orchid species that are grown from seed in culture produce green shoots after one or two years. I decided to investigate this by planting seeds in marked locations around the garden and in a local wood. The helleborine already grows in both places, and so to be satisfied that an orchid that appeared where I planted seed had come from my seed, many of my locations would need to produce one or more plants. What I failed to appreciate was the extremely low germination rate of Broad-leaved Helleborine, which I have subsequently discovered in yet another experiment. Altogether approximately 200 seeds were planted in each of 25 locations. However, based on the germination rate, the probability is that just one of these will produce a green plant, and thus it will be impossible to draw a conclusion.

Data collection and analysis

Clearly it is important that the starting hypothesis is carefully constructed, so that meaningful measurements can be made. Without proper guidance from the formulated plan the project can go off at a tangent. This may result in measurements of all sorts of interesting things which do not help to build up a picture of toxic material or population density or whatever other aspect of the biological landscape you are trying to study. Once you have decided what it is you are going to test, or look for, you need to ask the most important question, which is often skated over, namely how you intend to perform the analysis of your samples. Chemical analysis requires equipment, reagents and some knowledge. If your project involves this technical aspect, you may have to find another group or an academic department that can teach you the details of what you need and what to do. Even if you are using kits that are specifically designed to monitor, for example, the organic matter content of water, you will need some knowledge in order to get reliable and consistent results from them. This is also part of the joy of citizen science – you will be learning all sorts of new things that you never considered in the past.

As we saw in Chapter 3, citizen science is dogged by questions about the validity of the data collected. To obviate these problems, systems for data collection need to be in place from the very start. It should not be the case that test equipment is purchased without an extensive exploration of how it can be operated and in what form the results are presented. Even something as simple as a pH meter to measure the acidity of soil or water can cause problems if it is used in the field by untrained personnel or samples are stored for too long. Don't assume that technical equipment is essential – there may be a better and simpler method of measuring acidity and alkalinity, such as pH indicator papers. Even with a Universal Indicator, results can vary and are most definitely subjective, which sometimes makes them look random. When a group of individuals are asked to take measurements using a particular piece of equipment that requires any level of interpretation, the accuracy of the data will depend on the expectations of the participants and how clear they are about how interpretation works. Although overall accuracy may be compromised by a small amount, precision may vary widely from individual to individual. Of course, if protocols and methods are too complicated they will put people off, or the number of volunteers will decline as they become disillusioned and drift away.

An example of incorrect implementation of protocols would be recording a reading of 0 mm from a rain gauge that is damaged or for some other reason incapable of registering rainfall. To control errors and other problems with data it is necessary to be assiduous at the planning stage. This may need careful consideration of such things as the way that concentration is expressed, and to how many decimal places. If you are running a project that involves identification, such as Bee Watch, try to ensure that all of the citizen scientists are using the same identification guide so that any problems with specific identification are the same for all individuals, therefore affecting precision but hopefully not accuracy.

These and many other conditions that need to be controlled are best handled by a stable and consistent set of guidelines produced at the planning stage. Decide whether expert verification is needed, or if participant training will be sufficient to maintain data integrity across the whole project. It is interesting that although errors are naturally assumed to be errors of data honestly created, it has been suggested that some malicious data might be introduced deliberately. This may well be true, especially in cases where the study is particularly controversial or the results could be in direct conflict with a large, powerful business or with a government department that has a vested interest. The GLOBE Observer project, an international citizen science project designed to learn more about the environment and climate change, has training built into its reporting app which should help to reduce errors. Whatever aspect of the environment the citizen scientist is reporting – such as clouds, trees, land use or mosquito habitat – they must complete the training before they can start to report. Although this is probably not practicable for smaller DIY projects, when groups of people are involved it is always a good idea to provide some training in order to improve consistency of data gathering.

It is useful to create a recording form, especially if the project is a group affair, so that all of the required data are collected. Ideally the form will also summarise the protocol – another useful memory jogger for group members. The recording form has a long tradition. Daines Barrington, one of Gilbert White's correspondents in the collection of letters that form the *Natural History and Antiquities of Selbourne*, provided White with a standardised form for recording his phenological observations, although the term 'phenology' had not yet been coined. The form had columns for recording the date, temperature, pressure, weather, tree's first leaf, plant's first flower and so on. In a similar way the FIT Count form has tick boxes for the recorder's level of expertise and for habitat, flower cover, quadrat surrounds and weather conditions, as well as a tally count table for the insect visitors, and entries for date, time and target flower. Always keep a detailed contemporaneous note of your experiment or observations. Your memory is never as good as you think it is when you are young, and it is even worse than you think it is when you are old.

Once your data have been collected, the really interesting part of your project begins, because now you can start to analyse the data, and this can magically transform them into an understandable pattern of events. The methods of data analysis are many and varied, but when you do begin to look at your own data, if they are numerical then start by drawing a graph. This applies to any set of numbers, as the benefits of a pictorial representation of data are well worth the effort. It is sometimes surprising how easy it is to spot trends and changes very quickly from visual representations of data. To get the most from a graphical illustration it is important to choose the right method and style of graph or histogram. This will depend on whether values are independent or related, and also on what particular aspect of the data you want to emphasise.

Drawing graphs using commercially available software is easy, but it still remains your choice as to what style you use. Don't use a line to join up points of unrelated data – this is a common fault and can be very misleading. Graphs are tools of prediction, so as soon as you join up two points you are saying that the line between them tells you what happens between, say, two dates. In the case of seasonal rainfall, or the number of frosts in a winter, for example, the interim values cannot be

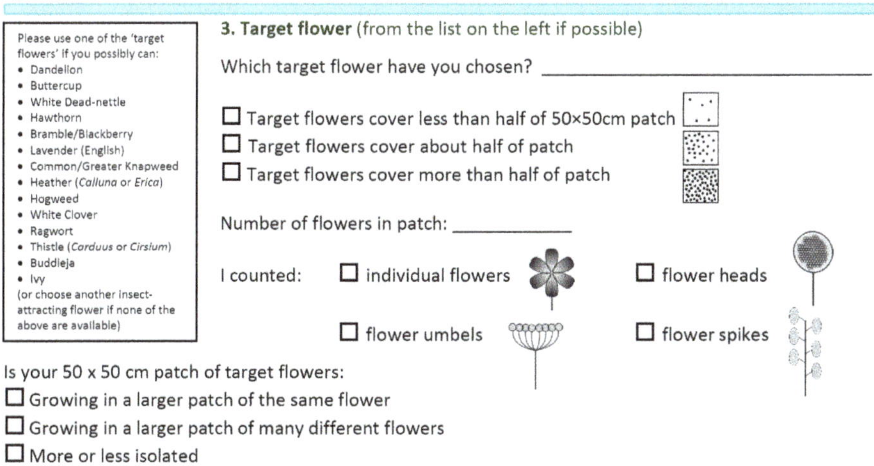

Please use one of the 'target flowers' if you possibly can:
- Dandelion
- Buttercup
- White Dead-nettle
- Hawthorn
- Bramble/Blackberry
- Lavender (English)
- Common/Greater Knapweed
- Heather (*Calluna* or *Erica*)
- Hogweed
- White Clover
- Ragwort
- Thistle (*Carduus* or *Cirsium*)
- Buddleja
- Ivy

(or choose another insect-attracting flower if none of the above are available)

3. Target flower (from the list on the left if possible)

Which target flower have you chosen? ____________________

☐ Target flowers cover less than half of 50×50cm patch
☐ Target flowers cover about half of patch
☐ Target flowers cover more than half of patch

Number of flowers in patch: __________

I counted: ☐ individual flowers ☐ flower heads ☐ flower umbels ☐ flower spikes

Is your 50 x 50 cm patch of target flowers:
☐ Growing in a larger patch of the same flower
☐ Growing in a larger patch of many different flowers
☐ More or less isolated

FIGURE 6.4 A section of the FIT Count recording form neatly demonstrating the utility of forms for both gathering consistent information and offering help to the recorder. (UK Pollinator Monitoring Scheme, 2024)

taken from the graph as they are unrelated to the rainfall that occurred before or afterwards. Also in relation to precision, make sure that you always label your axes. Nothing undermines a pictorial representation of data more than poorly or incorrectly labelled axes and the wrong style of graph. When you decide to construct a graph, consider what the data represent and think about whether a scatter graph, a line graph, a histogram or a pie chart is best for your purposes. When you have decided what style in the broad sense is most suitable, spend some time considering whether the data points should be joined up, and if so, whether this should be by a straight line or a curve. Sometimes what looks like a scatter graph of apparently random points can yield significant information by creating a regression line and its associated correlation coefficient. It is beyond the scope of this chapter to delve into the statistical processes and meanings of correlation, but it is well worth finding these things out before you start using them. Modern statistical packages will casually generate lines and correlations, which give a verisimilitude of truth to data, but this may mislead you into thinking that two things are linked when they are not.

What and where

That was a quick journey through the scientific method – all you need to do now is decide on a project. These will usually present themselves spontaneously once you start to observe nature in detail. Whatever it is that excites you, it is worth trying to develop that interest to see where it leads. One thing that we would always encourage individual citizen scientists to do at the very beginning of any project is to keep a notebook or journal. When I first took on a project to convert an overgrazed field back into a true acid grassland meadow, I started a notebook. This contained dated notes of all my observations, such as anything unusual, but also counts of things that were more ordinary, or frequently encountered. In this way I generated data from the baseline position of an excessively grazed field, as the slow biological changes that took place over a long-term scale of several years took place. The changes were slow because they were proceeding at their own pace, helped only by my management interventions, such as timed cutting of the hay. Looking back at the notebook it is quite apparent that there have been considerable changes over time, but at such a slow pace that they would not have been noticed without recourse to the notebook. A notebook is also useful for recording when new species appear, and how long the process of change has been occuring at the point when a new species arrives. For example, my notebook starts only with information about the growth of grass and Ragwort *Jacobaea vulgaris*, but then records the arrival of the first Marbled White *Melanargia galathea* butterflies and the increase in Rabbit *Oryctolagus cuniculus* activity as shown by the number of rabbit scrapes that were appearing. Later on this coincided with the appearance of mining bees and the abandoned rabbit holes being used by bumble bees for nest sites.

Of course there is nothing wrong with keeping a journal of observations just for the sheer joy of it. A handwritten journal is rewarding to keep and beautiful to look back on. One made by children with writing, paintings and drawings is magical for parents and grandparents, and for the children themselves when they have grown up. Some well-known naturalists have kept an annual journal recording their observations since their interest in nature was first sparked. For instance, Brett Westwood has been keeping a note of his observations since the mid-1970s and 'still enjoy[s] them as something sensational to read on the train'. However, and as I described for my meadow, your journal could include local change that you can watch, or animal behaviour that you can see, and this will almost certainly generate questions in your mind that can be developed into projects. While we are on the subject of notebooks and journals, it is best to use bound notebooks rather than loose-leaf ones, as the latter can lose pages, or the pages can easily get out of order.

Another useful source of inspiration can be participation in pre-existing citizen science projects, and adapting them for your own local area or species of your choice. For example, experience with PoMS and using FIT Count could lead into a study of different plant or insect species that are important in your locality. Margaret Percival, who we mentioned in Chapter 4 as an early employer of citizen scientists (the local girl guide troop), gained some fascinating ecological insights in a follow-up study, by close observation of the flowering of a range of wildflower species that were abundant in her area and regularly visited by the Honey Bee *Apis mellifera*. She noted that honey bees bring home hedgerow pollen from Blackberry *Rubus fruticosus* one hour earlier in the morning than meadow pollen from White Clover *Trifolium repens*. This is because bees do not collect pollen from dewy flowers, and the morning sun was drying the anthers and petals of the hedgerow flowers earlier than those of the meadow flowers.

At the other end of the scale from species diversity and ecology, inspiration may come from an area of scientific endeavour that used to be confined to technical laboratories and academic communities, but is now a legitimate part of citizen science. As we mentioned earlier, this is normally described under the general heading of 'DIY biology', and is usually associated with the handling and manipulation of DNA. This aspect of citizen science was very popular at the turn of the twentieth century but now seems to be attracting a diminishing number of citizen scientists. By its very nature, this activity has implications for both ethics and safety. As a result there are many regulations relating to this sort of experimentation that can be carried out on DNA and organisms, and these work well in established laboratories, but the information is out there for anyone to cultivate microorganisms or extract DNA. Although there are DIY biologists around who are looking at various genomes, their numbers are dwindling, probably because considerable technical expertise is required for many types of genetic manipulations that go beyond simple extraction of the genetic material. As we mentioned in Chapter 5 in relation to the Sussex Fungus Group, it is not impossible to develop the necessary technical expertise in a domestic kitchen or 'kitchen lab', but the production of negative results while the citizen scientist is developing the necessary skills can be frustrating.

A practical question you are likely to be faced with is where to conduct your project. If it is purely observational this is not so problematic, but if physical markers or other items of experimental equipment (e.g. nets, enclosures, trail cameras) are needed, these seem to attract passing members of the public who are often tempted to disturb them to see what is going on. In a public area you are at the whim of both the land owners and the general public. A friend of mine started what was going to be a long-term phenology project. He worked out an 800 m loop down through woodland, then alongside a riverside, past a pond, up a heath bank and finally along a hedgerow adjacent to a railway line. For the first 12 weeks of the annual flowering season he walked this route, noting the date and exact site of the first plant of each species to start flowering, and, to correlate with these, he obtained daily observations from a friend's nearby weather station (e.g. number of hours of sunshine, temperatures, rainfall). After a few years the local authority moved heavy machinery along the riverside path and trashed it, and railway workers cleared out the hedgerow, destroying most of his specimens. You should always seek permission from the landowner if want to use their land, so that they are at least aware of your activity, and they might also allow you to use a piece of land well away from public access.

If you are lucky you will have an area of your own to investigate, such as a meadow or a small piece of rough grassland. Also never overlook the possibilities of your own garden. Studies have shown that urban environments are just as species rich as rural ones, because intensive farming practices have dramatically reduced the biodiversity of the countryside. One citizen scientist has recorded almost 1,300 species in his garden in addition to all of the garden plants, and an RSPB survey found that urban environments are just as species rich as nature reserves and farmland. Of course, even on your own property you might not be entirely safe from interference. My grandson thought he could help with one of my pollination experiments by moving the quaint little covers I had put over flowers, and similarly John Box's children moved the coloured pins marking the position of mistletoe oaks on his large wall map of the UK to see if he would notice.

Another practical problem can be getting hold of the materials for your experiment. Sometimes you can construct these yourself. For example, an MSc student, Laura Brodie, was working from home and studying bumblebee foraging preferences. For her study she needed to mark individual bees and take measurements of their head size and tongue length. She caught the bees in glass jars, and made a restraining device from the top of a 6.5 cm plastic flower pot, a piece of mesh to cover it and the cap of an aerosol can with a piece of 1 mm graph paper stuck on to make the head measurements. A bee would be released from the glass jar into the mesh-covered flower pot, and the aerosol cap was then used to hold the bee against the mesh. Tongue length was measured by getting the bee to drink a honey solution from a plastic pipette marked with 1-mm graduations using a fine-tipped indelible pen, and the bees were marked with water-based Tipp-Ex of various colours. Her study site was a 400 m length of grassy verge along a local lane. If that is not DIY science, we do not know what is! These days commercial marking cages with similar design principles are available from natural history equipment suppliers for a very modest cost.

One final practical point, if you are setting up a group project, is to consider how to encourage your fellow volunteer citizen scientists to stay the course. Consider what is a reasonable commitment to ask for, and try to make the trigger for data collection neither so common that they get bored, nor so rare that they become frustrated. Most importantly, give feedback. It certainly encouraged me to continue to send in plant records when the county recorder always replied to my submissions to let me know how many new species in a monad I had found. And, of course, everybody likes to be mentioned in dispatches. Being named in reports, newsletters or acknowledgements is most satisfying.

We can summarise all of this as follows:

- Start from the beginning.
- Frame a question where you are interested in the answer.
- Decide what you are going to do with your results.
- Set up a testable hypothesis. This will focus your attention on the sort of data you need to collect.
- Decide how best to test your hypothesis.
- Decide if you are going to do this alone.
- What equipment will you need and where will the funding come from?
- How will you arrange and present your data so that they are easy to understand?
- How will you keep in touch with participants?
- How will you make sure that you minimise errors of all kinds?
- Don't be daunted by the apparent size of your project.
- Don't be discouraged by telling yourself that you are only an amateur – you are the backbone of citizen science.

Chapter 7

Get Out Your Notebooks and Start

As we have seen in earlier chapters, citizen science is an odd confection of many different projects and people, working alone or together. There are no simple answers as to what makes a citizen scientist. Indeed, one of the most interesting questions about citizen science is not why people become involved in large projects, but why individuals become so highly motivated that they will invest vast amounts of time and energy in becoming expert in a particular area of biology for the sheer joy and pleasure of it. Sometimes this enthusiasm may seem slightly eccentric, such as my delight in dung beetles, but that is only to outsiders. For everyone else this expertise

FIGURE 7.1 Citizen scientists in action: a transmitted light microscope with a true magnification of up to 1,000× is an invaluable tool for the citizen scientist, but only when properly used. It is worth getting expert help with this to learn the correct methods.

and enthusiasm is understood as being simultaneously of great intrinsic value and immense interest.

For some individual citizen scientists their fascination can start much later in life, although this does not make it any less important or the developed skill any less valuable. Although we are not going to suggest that everyone has the insight or knowledge of Charles Darwin or Gregor Mendel, these two individuals are examples of the range of age of onset. Charles Darwin was already interested in natural history at the age of eight, and this developed to the extent that by the time he was a student in Edinburgh he neglected his medical studies in order to spend time studying marine invertebrates. At the other end of the scale, Gregor Mendel studied philosophy and physics, and for much of his career as a working monk he was a teacher, primarily of physics. To then engage in plant breeding, producing his publishable results and starting the new science of genetics in his mid-thirties, was quite a feat. This author is more akin to Mendel, but only in the sense that my interest in plants developed later, in my first year at university, when I was looking down a microscope and watching cytoplasmic streaming made visible by the movement of chloroplasts around cells of the waterweed *Elodea*.

In the UK the macrofauna is greatly outnumbered by those species that only shine when viewed through a lens – the invertebrates. Meanwhile our flora sustains all of the animal species. Of course our large animals still need to be studied, as is clearly demonstrated by the work we described earlier involving Mr Holley's observations of the hares of Somerset. In whatever form and whatever the species, or group of plants or animals, you can make a contribution by just going out and recording what you see. We should just say here that, for the sake of brevity, we are using the terms plants and animals in a very casual way rather than an accurate one, since in modern science five kingdoms, not two, are recognised. Some of them you will be familiar with, others you may not be so aware of. Thus as well as the strictly plant and animal kingdoms, fungi are now regarded as a kingdom in their own right, as are the Protista, a hotch-potch group of Eukaryotes such as kelp, amoeba, slime moulds and protozoans all with a defined nucleus and the Monera, the group with no defined nucleus and no organelles, for instance bacteria. So, think in terms of the vast array of plants and animals that make their home together and make up an ecosystem. By becoming familiar with a specific group you will learn about the ecology of an area, which also means you will become an important part of the conservation of a habitat.

If you are planning to work on your own, this does not mean that you cannot access other people's knowledge and experience. This can be done via clubs and societies, either local examples such as wildlife trusts – every county has one – or on a broader basis by joining national or international societies. Don't think that you have to be an academic or already expert to join such societies; nowadays they have memberships for everyone, and the majority are very welcoming of beginners. The Royal Entomological Society, for instance, has several grades of membership from Associate Member through to Fellow, depending on your skill and ability. This arrangement is not unusual, and you may be very surprised by the range of subjects that have dedicated societies. So if you are interested in a specific species, or group

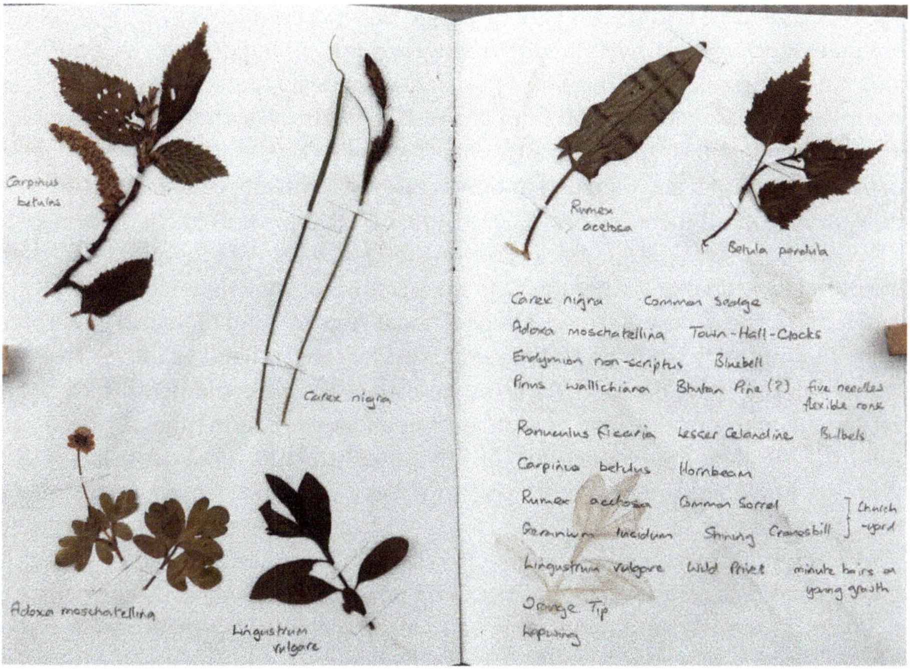

FIGURE 7.2 One of the author's notebooks begun when first learning the ropes for BSBI plant distribution recording.

of species have a look for a society that might hold some interest for you. Part of the reason for suggesting this is that it can be quite isolating if you are working on your own and have no contact with fellow enthusiasts. Making contact with likeminded people can also help in directing your curiosity so that you do not just repeat other people's mistakes. A large part of the joy of studying natural history comes from shared enthusiasm, and the thrill of finding rare or unusual species is much enhanced in the company of a group of friends.

It is almost inevitable that there will be mistakes when you start, these may be as simple as a problem with labelling specimens, or they may be more significant, such as errors of identification. The more complicated the system you are looking at, the more likely it is that there will be errors. So if you are just studying, say, a single group of insects, errors of identification may be few and quickly corrected; however, if you are looking at a whole ecosystem it may be a slower learning process to spot your own errors.

Magnifying your images: lenses and microscopes

In terms of studying specific groups of animals or plants, the more you become familiar with them the more detail you will want to look at. There is only so much that can be done with that essential piece of biological equipment, a hand lens, so you may want to progress towards a microscope. For the biologist, light microscopes come in two broad forms: the dissecting microscope and the transmitted light microscope. The dissecting microscope is deceptively titled, as very few people use them for dissecting, even though they were originally designed for that purpose and still can be employed that way. The plant or animal of interest is placed under the microscope and illuminated from above. The depth of field, that is the amount in focus at any time, is relatively large with these microscopes. Having a reasonable depth of focus is ideal for looking at details needed for identification which are not truly microscopic but are very small. Dissecting microscopes are easy to use and with a few accoutrements, such as fine probes and forceps, will suffice for most large-scale biological investigations. These microscopes can be bought quite cheaply, but be aware that a cheap new one may well have some very irritating flaws and be surprisingly poor optically compared with a good quality second-hand example. An even cheaper device is the digital USB microscope attached to a computer, these can be very useful with practice, but resolution of fine features can be poor. Now, if you want to look at far more detail than just the gross structure of an organism, if you want to delve down into the very cells that make it up, you will need a transmitted light microscope.

Transmitted light microscopes are often badly maintained and badly used, so it is worth finding a course or books (for example *Worlds within Worlds: An Introduction to Microscopes* by Bruno Kremer) that can help. For this we would encourage you to get in touch with the Royal Microscopical Society, and to have a look at the virtual microscopy tutorials and experiments at the Open University's web-based OpenScience Laboratory. It is most unlikely that you will have been shown the correct way to set up a microscope at school, as it can be quite an intricate process.

FIGURE 7.3 A dissecting microscope is very useful when dealing with small details and avoids the eye strain associated with hand lenses.

A good quality microscope can operate at the limits of optical resolution, a limit set by physics. Nonetheless, used properly, resolution of subcellular organelles, such as chromosomes is perfectly possible, and chromosome number and structure can tell us much about evolution within a group. Although there is a lot that can be said about microscopes, there is one thing to bear in mind when looking to buy one: adverts that claim a magnification beyond about ×1,000 are leading you up the garden path. Anything over this figure can only be claimed to be a magnification, in the same way that projecting an image onto a wall will enlarge it – but no more detail will be forthcoming from the image.

There is something that needs to be considered when embarking on a biological study using a transmitted light microscope: the stains and dyes that will be needed along with glass microslides for mounting your specimens. Although each one is not in itself expensive, the additional costs will add up. This little bit about microscopes may seem of minor relevance overall, but detail needs magnification and some detail is invisible to the naked eye without it. Detail is also invisible with a microscope unless you increase the contrast within your specimen.

All this brings something else to mind regarding the citizen science sole operator: the fact that you will be learning all manner of ancillary stuff, which involves new forms of mental agility. This in itself is an excellent thing for anyone, but if by chance or design you have survived your years of work and are now retired, it is said to be even more so. There is one other thing that sometimes gets forgotten but should be at the front of your mind: it will be fun. We do not say this flippantly, whatever you take on, whatever you decide to investigate, it should be an enjoyable activity for your sake as well as for the environment.

Getting Your Data to a Wider Audience

From all of this, it is obvious that there are things that you need to think about when considering an individual project. This is true whether you are just interested in going out recording species you see, or you have spotted something that you want to know the answer to but have been unable to find anyone or anywhere that seems to know. The first thing is motivation. Anything you undertake on your own is going to have to be self-motivated and this includes citizen science projects. Part of the motivation for a citizen scientist is finding out something useful and distributing this information beyond their small group. The most obvious way in which to do this is to publish your results in a scientific journal. As we have seen in earlier chapters, it is not always easy to get scientific results into print if you do not have an academic affiliation. While this has made life difficult for some citizen scientists, the thinking behind it is simple. Journal editors receive large numbers of submissions for publication and of those, the vast majority of the misguided and agenda-led papers come from private addresses with no scientific credentials. This has resulted in all private correspondence being tarred with the same brush of doubt. While this may seem a rather heavy-handed blanket response, it should be realised that most scientific journals are edited by volunteers who are usually fully employed academics. So, to save the editors from being overworked, they make an arbitrary decision on private addresses. Although this does mean that publishing your results can be difficult, there are ways around the problem. One way is to gain the help and assistance of an academic partner and the other is to publish in journals that do not have preconceived ideas based on a private postal address. In the twenty-first century it is more likely to be an email address that does not end with .ac.uk or .edu that tips the balance between ignored and read. What you may be looking for are local society journals – after all, Gregor Mendel, the founder of modern genetics, published his work in the journal of the Brünn Natural History Society. There are many other suitable places for publication and by searching and looking at the contents you will be able to judge which one is most suitable for your work. You may imagine that local journals have no staying power, but any published material such as a local or amateur journal in the UK will be kept at the countries six legal deposit libraries, which means they can be viewed at anytime in the future by anybody. These days of course, most society journals and newsletters are published online so they will remain available for at least as long as the society continues to exist.

If you do publish your results, remember that simply reporting an observation alone, as new and exciting as it may be, is not really science. Close observation, such as Percival's work on honey bees and wildflowers, will almost always turn up some novel observations but what is important is the interpretation of these in the light of the current understanding of ecology, or whatever science the observation pertains to. Even some papers coming from academic origins constitute what can be broadly described as natural history, being purely descriptive and of little scientific content. The problem is that in the twenty-first century there is professional pressure in science to publish as often and as much as possible. Although such papers can be interesting they do not generally add anything to the sum total of knowledge. Science works by fitting observation to reality and if it does not fit, say why, or find out why, create a hypothesis to test. When done in that way and the results published, then it may make a significant difference to our knowledge by allowing predictions to be made about similar situations. Purely observational papers are not valueless, however. They can be an excellent stimulus for others, including professionals, to take a look.

If your citizen science project was, say, investigating river pollution and you are publishing with a view to affecting public opinion or gaining public engagement in your cause, then a scientific paper in a learned journal is probably not going to have the desired effect. Research and experience shows that facts do not necessarily change minds even where the audience has an understanding of the subject, and you should bear in mind that the majority of the general public are not well versed in science. The brain naturally tends to rely on assumptions and preconceived notions when faced with large or complex data sets, and humans are more prone to being convinced by individual stories rather than statistics and probability. One can see this with the Post Office scandal of the early twenty-first century. The facts were out in detailed radio documentaries and press reports for many years but it took a TV drama relating the devastating impact on individual cases to galvanise public opinion. So, in this situation, as well as your scientific paper, which it is still important to refer to, you need an article with a completely different approach in a more widely read publication. If the issue you are looking at is local, then the local press is a good place to start.

Although it may seem counterintuitive, when you are working on a project singlehandedly the question of equipment does not generally arise. This is for two reasons, the first is that it is unlikely that you will need a lot of big and expensive equipment to get started. The second reason is that most citizen science projects of this type will have originated from a hobby, which in itself usually accumulates sundry bits and pieces which will seamlessly move into being used for your citizen science project. If you do find that you need something specialised for your project, try and borrow it. Of course, if it is going to be needed on a regular basis, buying your own may be the best answer. You may think that borrowing a piece of specialist equipment is going to be almost impossible, but it may be worth your while seeking out likeminded individuals who already have the wanted bit of kit, or may want to share the cost. For example, the West Midlands Ringing Group found that a shared thermal imaging camera was sufficient for all their members. You may want to meet up with such people anyway, as a way of exchanging information and ideas which will help you in clarifying your own thoughts and concepts.

Clubs and Societies

If your interest lies with birds, there are local ornithology clubs and in the UK every county has a Wildlife Trust that will often also have an amazing range of specialist interest groups within it. On a more national level there is the RSPB and the British Trust for Ornithology (BTO). It is the BTO that runs and organises the UK bird ringing scheme, for which they also supply the legally required bird ringing licences, and run the training courses that the licences require.

For mammals in the UK, there is the Mammal Society or the Bat Conservation Trust, which of course will bring highly specialist knowledge to the table. There are also many local groups that have a particular interest, such as deer study groups in most of the larger forests. Other animal groups, such as amphibians and reptiles, are helped by work initiated by the RSPB as well as Amphibian and Reptile Conservation. Perhaps oddly, since we have no native breeding chelonians in the UK, there is a thriving conservation charity the British Chelonia Group, helping tortoises, terrapins and turtles. I should add here that we do have a considerable visiting population of turtles around our shores, but they do not breed here. Turtles are occasionally stranded and three species are regularly reported in British waters, Leatherback, Kemp's Ridley and Green Turtle, nearly always as juveniles. We know this because as well as the help of the Coast Guard around the UK shores, citizen scientists were encouraged to report sightings or stranded specimens to the Natural History Museum in London. In 1967 the British Museum (Natural History) as it was called at the time published a booklet that encouraged those finding stranded specimens to send them 'to the Museum alive, packed in a strong case or ventilated box by passenger train' (Brongersma 1967). A telegram should also be sent informing the museum of its imminent arrival. A turtle in the post, whatever next?

Among the smaller animals, invertebrates are championed by Buglife, and with a more focussed target, insects are at the centre of the Royal Entomological Society (RES), the Amateur Entomological Society (AES) and a whole raft of specialist groups such as the Dipterists Forum. If you are interested in some of our more poorly researched groups of invertebrates, the Field Studies Council produce a very wide range of dichotomous keys for identification. They have managed to publish identification guides for such diverse groups as earthworms, freshwater snails, Sphagnum mosses and intertidal marine isopods. If you have a project, or just a background interest in a group for which there seems to be no easy way of identifying to family or species level, look at the available keys to identification and you will be surprised at what you can find.

In the plant world there are societies for just about every rank conceivable – for higher plants and ferns there is the BSBI, the Wild Flower Society and Plantlife, for ferns there is the British Pteridological Society, for mosses and liverworts the British Bryological Society, and for lichens the British Lichen Society. There is even a British Plant Gall Society which crosses ranks because it is interested in both the plants that the galls are on and the insects that create them. Nearly all of these societies run field meetings, which is the best way of learning, and also produce specialist keys for identifying particularly difficult groups.

Methods of Identifying Species

On the subject of identifying different species, it is relatively easy in the UK to identify large animals by using one of the many available guides, illustrated either with photographs or drawings. This reflects the small number of different species of this type we have in Britain. For smaller organisms, both animal and plant, this is not so easy. Even if field guides exist, in many cases they are of little help. The reason for this is straightforward: the differentiation between species can be very small indeed, so small it needs detailed examination to be certain of what you have. The way around this is by using a dichotomous key, as they will help you to identify with some certainty the plant or animal you have in front of you. It does take both time and practice to do this as the terms and descriptions can be confusing when you start, but the outcome is well worth the effort. It takes a lot of determination and effort by the authors to create a reliable key and many of them go through extensive testing before they are unleashed on the public – as one of us can testify.

So often, identification is regarded as simply a number collecting exercise, collecting species for its own sake, but it has far more to it than that. It has been suggested over the past years that taxonomy – the science of describing and naming species – is in decline. It is true that formal courses on the subject are few and if it is taught at all it tends to be as a small part of a wider biology degree. Unfortunately, the lack of formal courses is reflected in the lack of interest in taxonomy among scientists as a whole. This is especially seen in the rather more reductionist subjects, such as molecular biology where subjectivity and opinion stands for little. What this has resulted in is a gradual decline in numbers of professional taxonomists. By way of illustration, only a handful of the several hundred public herbaria in the UK are curated any longer by keepers who are plant taxonomists, although some of this is no doubt due to lack of funding as well as of taxonomists. Hopefully this situation is slowly on the turn as citizen science projects have started to throw up interesting statistics regarding biodiversity, simply by identifying trapped and caught species.

It is easy to assume that highly populated areas such as Europe, which have a long tradition of natural history, have had all the different plant and animal species named long ago. Strangely this does not seem to be the case. By collecting data from journals that publish descriptions of new species, it has been possible to show that over the last 70 or so years, about 770 new species have been described every year. It would be easy to imagine that this includes a lot of tiny and miniscule microbes, but this figure is for multicellular organisms only. Taken in comparison with global figures, this may seem a rather small number, there being around 17,000 newly described species found in the rest of the world annually. Back in Europe, the most remarkable thing about all these new discoveries is not that they are found in such a high density living space, but that more than 60% of them are described by non-professional taxonomists, citizens scientists. We use the term non-professional rather than amateur for a reason; the former term denotes unpaid, the latter less than professional.

By taking a specific group and developing the skills in identification, the citizen scientist can easily come to the point where they are able to say that a specimen represents a new species. While it is true that most vertebrate species, mainly birds and mammals, are fully described, in recent years freshwater fish new to science have been described in Europe. It is, however, in among the smaller species, the invertebrates, that non-professional taxonomists shine and for some of us it is in these groups that the most astonishing things are found. For most of these organisms, even though they are clearly visible with the naked eye, a dissecting microscope is useful for detailed examination. If you are interested in the even smaller, then microscopy becomes an essential part of your developing skills. Microscopy was an essential part of the work of Alfred Kahn (1877–1946) a citizen scientist who can be said to have taken over a field of study and become a leading expert on it. His field of expertise were the ciliate protozoa, small single celled organisms. This was an interest he developed while a teacher in Germany. Cilliate protozoa are a group of single celled organisms found in water and they very definitely need a microscope to be seen clearly. During his studies Kahn not only described around 700 new species, but extended his knowledge to be able to rearrange the group with 57 new genera and 17 new families. That really does demonstrate a deep understanding of the relationships within his study group, but also how taxonomy works at all levels to help understand evolutionary relationships.

More recently developed techniques in taxonomy, such as DNA analysis, are not always straightforward for non-professionals to access. It is very easy to think that it is now possible to identify any species from a DNA sample, but this is not always true. As with many such processes there is an element of interpretation of the results required when looking at DNA profiles. Needless to say, creating DNA profiles can be difficult anyway for the amateur with limited resources, though not impossible. With DNA analysis there can be a disparity between the technologically able, the molecular biologist, and the taxonomically aware, the taxonomist. This can arise because differences in DNA said to define a species may be at odds with the taxonomists knowledge of the species being investigated. Part of the perceived value of defining a species by its DNA is that a barcode can be assigned to a species, but this gives little away to the most important part of taxonomy which is the relationship between species. In times past taxonomists were often described as either 'lumpers' or 'splitters'. Lumpers were those who reduced the number of species by grouping two or more controversial species together and splitters were those taxonomists that did exactly the opposite by dividing a species into two or more different ones. It is now possible to produce a barcode for any given species, but whether DNA barcoding will affect this aspect of taxonomy remains to be seen. Barcoding is carried out by the Biodiversity Institute of Ontario which is part of the University of Guelph in Canada and is generally beyond the reach of amateur collectors.

In a similar way to DNA being seen as a universal definer of species, so, too, is supplying access to digital databases of photographs. This is unfortunately not the case; for example, many insect species can only be defined by microscopical examination, often of the genitalia, so a simple photograph will only give a superficial identification, especially in complicated species groups. Needless to say, the more you become versed in your subject group the more likely you are to accurately identify rare species and species that have not yet been described and named. The moral here is that getting out and getting familiar with a group of species first hand cannot be overstated in importance.

As we have seen, in times past the lone citizen scientist would have been the normal state rather than the exception, but in truth such individuals are now rare. In terms of field activity, it is the grand mass of volunteers who are powering ecological and conservation studies. Nonetheless, you can still make a useful contribution from what you can study in your own back garden in your own way and your own time, resulting in a very satisfying sense of achievement. A very real modern contributor to our knowledge of palaeontology is a lone citizen scientist, Heather Middleton. We have all heard of Mary Anning (1799–1847) who collected fossils around Lyme Regis on the south coast, at a time when palaeontology was in its infancy. So we would assume that two hundred years later the state of knowledge regarding the fossil beds of Britain would have increased beyond the power of a citizen scientist to be of any significance. Not so, as clearly demonstrated by Heather Middleton. After a lifetime as a teacher she retired at 60 on the south coast, close to the small tourist town of Weymouth, where she started collecting fossils. After 20 years she had accumulated more than 2,000 fossils and a perfect attitude, epitomised by 'Retirement is really an opportunity to explore new interests – if you're healthy.'

The collecting that Heather Middleton started was never intended as more than a personal activity, but it has turned into a citizen science project. As a consequence it has produced material of immense value to professional palaeontologists who have collaborated on publications and used the private collection for their own research. The very nature of what such a project should be was summed up when she commented: 'What immense fun I've had. It's fun, for goodness sake.' One of the things that stands out among lone citizen scientists is that they often tend to be collectors of natural objects, just as Heather Middleton did in her childhood. This may, as in the authors of this book, develop into careers in science, or it may be sidelined for later. Either way, thinking about what you did as a child can help explain that urge you have to look under logs, photograph the living world and study as an adult of any age.

Whatever you are studying – be it a species, a population or an ecology – there will be things that you don't know, but are already known. Always read around a subject to get a feel for how much has already been investigated, you may find a lot about your chosen area of investigation that way. So find out as much as you can as it will hold you in good stead when you come to look at your own data in detail. These days there is a tremendous amount of good scientific literature available online. Search using Google Scholar and check out sites such as ResearchGate, JSTOR, ETHOS, Oxford Academic, Academia.edu, the Wiley Online Library, the Hathi Trust and MDPI.

If you are still keen on a singlehanded study, it is worth considering some spectacular achievements by individuals. Guy Callendar (1898–1964) is one such citizen scientist, he came from an academic family and he was professionally involved in studies of steam and pressure for industrial companies with regard to turbines. Later in his professional career he turned his attention to batteries and fuel cells. Besides this, and some may claim, more importantly for the planet, Callendar was also an amateur climatologist. It was in this capacity that he used temperature measurements from the nineteenth century and matched it to atmospheric carbon dioxide (CO_2) concentration for the same period. From the approach of looking at what were assumed to be two apparently disparate and independent variables, by 1938 he had not only seen a steady increase in the temperature of the planet's landmasses, but had associated the two together. By correlating the two variables, he produced estimates of temperature rise which turned out to be accurate. This gives us an estimate of global temperature rise of 2% for a doubling of CO_2 concentration. Initially, professional meteorologists tended to think the association of carbon dioxide and temperature rise was coincidental, not causal. Callendar's continued publication of accumulated data helped to convince many of the otherwise sceptical of the validity of his ideas of a changing climate. It should be said here that in many ways scepticism of his results was not surprising since in 2022 carbon dioxide still only formed a tiny proportion of the atmosphere, just 0.04% (very strange but true). So until it was understood how this molecule can affect climate and temperature, the association seemed far fetched. Individual citizen scientists can have many changes in direction and we can never know what a particularly clear-thinking citizen scientist may come up with in the future.

Should you be more inclined towards group activity, and are still determined to be involved with citizen science, then there are two routes you can take. Both of these assume that you are highly motivated by some aspect of the world in which we live. Environmental investigations of all sorts, whether it is counts of individual species or numbers in a population, measures of land use or river pollution, these are all ideally suited to group investigations. So given that you are motivated to help the planet in some way, the first route you can take would be to set up your own group, starting from scratch on a completely new project. The second route is to join an existing project. The second route, by its very nature is the easiest to become involved in, but it may not suit you or be as satisfactory. On the other hand, this can be a way of learning the practicalities of setting up and running a group. This may be a very good prelude to you spotting a question that you want answered and founding a group of likeminded people to take on the challenge.

If you plan on joining an already-existing group, even if it is newly formed, there are a few questions you should ask so as not to waste your time and effort. This is the case regardless of what the project is: a citizen science project is no different to any other science project, it has to be approached in the right way, or the results will be unrepeatable and consequently of little value. So, be sure to frame some appropriate questions that can be asked to clarify the project. The most important thing to ascertain is whether the project is asking a question, and more than that, whether it is asking a question that can be answered. This is important because although simply collecting data can be of value, as in phenology studies, it can be a long term activity that would normally be associated with a larger project in its own right.

The question being asked by the group you want to join may be, for example, about a species distribution, or possibly a population density. Alternatively, it may be trying to ascertain the level of pollution in a lake or river. These are the questions that you may find have already been answered by projects run by societies like the RSPB, or anti-pollution groups. Under these circumstances you may not wish to reinvent the wheel by repeating other people's measurements. If the information is already out there and you repeat the measurements what is it going to be used for? An intrinsic part of citizen science projects is dissemination of resulting data, so you may find you are in a group where feelings run high, fanned by frustration at their own data not showing what they want it to.

If you intend starting a group yourself, spend some time deciding on what the aims of the project are. But it is worth being aware that just because you kick-off the project it does not mean that you will, or should, remain in control of it. This is especially the case where everyone has an opinion about the project, so it is important to remain flexible in attitude so that something comes out of the study.

When framing your aims, be as specific as possible. If you are concerned about pollution in a river or lake, one aim may be to measure it, but that will not do anything about the pollution, or your concerns. So state in your original plan what you intend to do with the data and the information it yields. If you find an adverse result it must be recorded, but it is equally important if you do not find any pollution, and you are confident with the results of your analysis, that you say so. Many local papers and local radio stations as well as Facebook groups and other websites will be delighted to trumpet a positive outcome just as much as a negative one, for the simple reason that if your river is clean everyone will be happy.

Something that does need to be addressed in all citizen science projects, whether they are conducted by an individual or by a group, is the failure of the task, or at least part of the task. There are many reasons that a project might fail, but it comes down to either the scientific component or the management component, this latter includes dealing with individuals and framing questions. Although failures rarely cause the complete collapse of the project, they can make it difficult to take the results at face value. It should be remembered that failure is a normal part of the scientific endeavour, regardless of who is involved. This is well worth emphasising for the volunteers who may be unaware of how scientific enquiry works, This is especially so when it comes to how long it can take for the most complex tasks to be fully understood, not to mention how long the associated techniques may take to start working in a useful and consistent manner.

The investigation of failed citizen science projects is a rare phenomenon and is usually carried out as a sociological study. Where a thorough look at the various projects has been undertaken, a consistent story starts to appear. Usually the failure is not of the entire project, but more of retention of skilled citizen scientists for projects where the participants need to learn specific methods and techniques. Before considering the categories and areas which were reported as points of trouble, when setting up a project it is advisable to consider the one area that the published papers never mentioned, but is regularly seen. This is that people are volunteering for these projects and there will be a natural wastage. Participants may just lose interest in the idea, or find they have other commitments that take them away from the geographical area or reduce their available time.

Published problem areas with citizen science projects all seem to be about some form of difficulty in dealing with people. The problems described were time management, methodology, complexity of the task and the biggest stumbling block of them all, communication. However, with regard to communication, one aspect that was always regarded as a success came when the results were in. If they were published and disseminated among the participants, regardless of whether the results were seen as success or failure, good or bad, those involved were happy with the outcome.

Most major points of failure in citizen science projects seem to originate from the execution of the required task. This failure seems to have nearly always been due to lack of communication. This resulted in a consequent lack of ability of the participants to work effectively with whatever form of data-collecting equipment was provided. Failures of tools or measuring devices of all sorts was associated with inadequate training, or sometimes the equipment being just too complicated to use. This is in contrast to activities that required people to do things of a practical nature and then observing what changes were seen in biodiversity. This could be as simple as collecting samples or digging ditches and seeing what turned up.

A citizen science project looking at bees in Chicago designed to garner details on species and distribution of pollinating bees in and around the city failed, as it did not appear to have engaged with the participants. Reading the final report, there does seem to have been an element of the data collectors being patronised and – more important even than that – the participants had to actively engage with digital platforms. This was the only way for them to gain information about the progress of the project and the results from their activities and of the completed project. This highlights what tends to be forgotten in citizen science projects: participants are volunteering their time and effort to help on the practical side, so it is incumbent on the organisers to make sure that they communicate with participants and keep them up to date. It should not be the job of the participants to have to find out for themselves what the results are. It all comes down to this; people will feel much more personally involved if they are given information rather than having to go and find it for themselves.

This idea of engaging with participants has created all manner of suggestions as to how it can be done effectively, from working on projects with social scientists, to engaging science educators to help in the process. But rarely is it said that the organisers of the project should simply give out information as a routine part of being in charge of the citizen science process. This is instead of expecting participants to log-on to a system and search out information about the project in which they are not only involved, but are a pivotal part.

One extremely important aspect of citizen science in biology is very simple, but almost impossible to nail down and quantify. This is the benefit to the participants of being in nature. This may sound somewhat nebulous, but we can, at least in part, describe some of the factors and events that make a difference to people's mood when in the countryside. There does seem to be a specific correlation with health and perceived wellbeing with access to nature, but it goes further than that by being noticeably enhanced by increased biodiversity. To complicate this, it does not seem in those studies that have been carried out, that the length of time or number of visits to open green spaces explain the benefits, regardless of biodiversity. However, what does help to explain the benefits is the attitude of the visitor to the countryside, put another way, their level of connection to the natural world. This attitude is, of course, a very subjective thing, but if an individual feels connected to nature, so they will reap the benefits of being outside in a species-rich environment. Citizen science projects are extremely good at this process of getting people out, but be aware that the sense of being connected to nature can be undermined by citizen projects that have a high technological component. Projects of this type are easy to imagine, they are such things as testing for water pollution by chemical analysis or weather station monitoring, where data is primarily numerical and is collected by participants and sent off to a data station. At the other end of the process, testing for water pollution by checking the plants and animals in a watercourse, will bring the participants into very close proximity to nature and all its benefits.

It has been suggested in many articles that there is a problem of few younger people being engaged in citizen science, well most of the non-academics among you will be only too aware of the reason for this. Throughout the time that sociology researchers are teeth gnashing over lack of youthful engagement, the youthful are engaged in trying to construct a career and associated social activities; just not enough time is available for everything. So do not fret if all your compatriot citizen scientists seem middle aged and over. This is similar to older professional footballers not fretting that all their compatriots are young. Some things fit better within a specific chronology and citizen science will always benefit from background knowledge and experience that comes with age and a wider outlook.

Citizen science has grown exponentially this century and there is no doubt that it will continue to do so. Mainstream science has come, or perhaps it is fairer to say is coming, to accept its validity and importance, particularly in ecology and conservation, and, indeed, to respect and value the expertise of many non-professionals. Why not join in? As you will have seen from the brief list at the end of the last chapter, doing your own science is not complicated. What is complicated is raising and maintaining motivation to undertake what may turn out to be a long and complicated study. The best way is to remain focussed on your target, but be flexible in your outlook so that any new and intriguing results that point you in a different direction can be accommodated, allowing you to change emphasis or direction and move the project in a different and more productive direction. There is one thing you should definitely aim for and that is to enjoy it. So get out your notebooks and give it a go.

Bibliography

Allen, D. E. (1976) *The Naturalist in Britain: A Social History.* Pelican Books.

Andersson-Sundén, E., Gustavsson, C., Hjalmarsson, A., Jacewicz, M., Lantz, M., Marciniewski, L., Ziemann, V., Barker, A., Lundén, K. (2019) Citizen Science and Radioactivity. *Nuclear Physics News* 29(2): 25–28. https://doi.org/10.1080/10619127.2019.1603559

Aono, Y. and Kazui, K. (2008) Phenological data series of cherry tree flowering in Kyoto, Japan, and its application to reconstruction of springtime temperatures since the 9th century. *International Journal of Climatology* 28: 905–914. https://doi.org/10.1002/joc.1594

Blamey, M., Fitter, A. and Fitter, R. (2003) *Wild Flowers of Britain and Ireland.* A and C Black, London.

Brodey, B. (1996) Bumblebee Foraging Preferences: Differences Between Species and Individuals. Unpublished BSc thesis, University of Aberdeen. Available at: https://citeseerx.ist.psu.edu/document?repid=rep1&type=pdf&doi=275c44c2f871e2046ba314fabbc92c7cd991c86d

Brongersma, L.D. (1967) *British Turtles.* The British Museum(Natural History), London.

Cooper, D. (1839) *Proceedings of the Botanical Society of London* pp. 3–6. Available at: https://www.biodiversitylibrary.org/item/189042

Corbet, P.S., Longfield, C. and Moore, N.W. (1960) *Dragonflies.* New Naturalist 41, Collins, London.

de Grey, A.D.N.J. (1999) *The Mitochondrial Free Radical Theory of Aging.* R.G. Landes Company. Available at: http://www.r-5.org/files/books/medicine/senescence/oxy-stress/Aubrey_D_N_J_de_Grey-The_Mitochondrial_Theory_of_Aging-EN.pdf

Falk, S., Foster, G., Comont, R., Conroy, J., Bostock, H., Salisbury, A. et al. (2019) Evaluating the ability of citizen scientists to identify bumblebee (*Bombus*) species. *PLOS ONE* 14(6): e0218614. https://doi.org/10.1371/journal.pone.0218614

Fitter, A.H. And Fitter, R. (2002) Rapid Changes in Flowering Time in British Plants. *Science* 296 (5573) 1689–91.

Fitter, R. (1945) *London's Natural History* New Naturalist Number 3, Collins, London.

Fitter, R. (1952) *Collins Pocket Guide to British Birds.* Collins, London

Holley, A.J.F. and Greenwood, P.J.T (1984) The Myth of the Mad March Hare. *Nature* 309: 549–550. https://doi.org/10.1038/309549a0

Hunter, J., Alabri, A., van Ingen, C. (2013) Assessing the quality and trustworthiness of citizen science data. *Concurrency and Computation: Practice and Experience* 25: 454–466. https://doi.org/10.1002/cpe.2923

Kerson, R. (1989). Lab for the environment. *Technology Review* 92(1): 11–12.

Kremer, B. (2024). *Worlds within Worlds: An Introduction to Microscopes.* Pelagic Publishing.

Kuhn, T. (1996) *Structure of Scientific Revolutions.* Third Edition. University of Chicago Press, USA. https://doi.org/10.7208/chicago/9780226458106.001.0001

Lack, D, (1943) *The Life of the Robin.* H. F. & G. Witherby Ltd. London.

Longfield, C. (1937) *The Dragonflies of the British Isles.* Frederick Warnes and Co. Ltd., London

Lovelock, J. (2000) *Homage to Gaia.* Oxford University Press.

Leopold, A. (1949) *A Sand County Almanac and Sketches Here and There.* Oxford University Press.

McClintock, D. and Fitter, R (1956) *Collins Pocket Guide to Wild Flowers.* Collins, London.

Mendel, G. (1865) Versuche über Pflanzen-hybriden. *Verhandlungen naturforschender Verein in Brünn,* Abhandlungen, iv, 3–47. Dated 1865, published 1866. Translated at the behest of the Royal Horticultural Society with minor changes by William Bateson as Mendel, G. 1901. Experiments in Plant hybridisation. *Journal of the Royal Horticultural Society* 26: 1–32.

Morgan, D. (2021) Dull flowers and dazzling insects – the story of pollination in one of our less glamorous orchids. *British Wildlife* 33: 118–124.

Pocock, M.J.O. and Evans, D.M. (2014) The Success of the Horse-Chestnut Leaf-Miner, *Cameraria ohridella,* in the UK Revealed with Hypothesis-Led Citizen Science. *PLOS ONE* 9(1): e86226. https://doi.org/10.1371/journal.pone.0086226

Porter, R. (1978) Gentlemen and Geology: The Emergence of a Scientific Career, 1660–1920. *The Historical Journal* 21: 809–836. https://doi.org/10.1017/S0018246X78000024

Potter, B. (1989). The Journal of Beatrix Potter 1881–1897 Complete Edition. Translated from her coded writings by Leslie Linder. Fredrick Warne, London.

Siemensma, F. *Microworld: World of Amoeboid Organisms.* https://arcella.nl/

Squires, T.M., Yuda, P., Akbar, P.G., Collar, N.J., Devenish, C., Taufiqurrahman, I., (2021) Citizen science rapidly delivers extensive distribution data for birds in a key tropical biodiversity area. *Global Ecology and Conservation* 28. https://doi.org/10.1016/j.gecco.2021.e01680

Stubbs, A., Falk, S. (1983) *British Hoverflies an Illustrated Identification Guide.* British Entomological and Natural History Society, London.

The Science of Citizen Science (2021) eds. Vohland, K., Land-Zandstra, A., Ceccaroni, L., Lemmens, R., Perelló, J., Ponti, M., Samson, R. and Wagenknecht, K. Springer.

Toomey, A. H. (2023) Why facts don't change minds: Insights from cognitive science for the improved communication of conservation research. *Biological Conservation* 278. https://doi.org/10.1016/j.biocon.2022.109886

Voelcker, E. and Clauß, S. *Wunderkanone. Exploring the Microscopic World* http://www.wunderkanone.de/

Wall W.J. 1999. *Sexing the Parrot: Changing The World With DNA.* Cassell, London.

Watson, H. (1832). Outlines of the Distribution of British Plants, Edinburgh, Private publication.

Watson, H. (1847–1852). *Cybele Britannica: or British Plants and their Geographical Relations.* Longman and Co., London.

Watson, H. (1873) *Topographical Botany.* Bernard Quaritch, London.

Welz A. (2019) Listening to Nature: The Emerging Field of Bioacoustics. *Yale Environment 360.* Yale School of the Environment. Available at: https://e360.yale.edu/features/listening-to-nature-the-emerging-field-of-bioacoustics

West Midlands Ringing Group (2020). Thermal Bird Surveys. https://www.youtube.com/watch?v=Yb5xjgyXqTY

White, G. (1975) *Garden Kalendar 1751–1773.* The Scholar Press, London.

White, G. (1788–9) *The Natural History of Selborne* Penguin reprint 1977. Penguin Books, London.

Wibowo, W.K., Winarni, N.L, Yanuar, A. and Marsden, S.J. (2021) Citizen science rapidly delivers extensive distribution data for birds in a key tropical biodiversity area. *Global Ecology and Conservation* 28: e01680. https://doi.org/10.1016/j.gecco.2021.e01680

Woods, Mr. (1839) *Proceedings of the Botanical Society of London* pp. 54–55. https://www.biodiversitylibrary.org/item/189042

Index

Entries in *italics* pertain to figures.

C

D

E

F

G

H

I

J

K

L

M

N

O

P

www.ingramcontent.com/pod-product-compliance
Ingram Content Group UK Ltd.
Pitfield, Milton Keynes, MK11 3LW, UK
UKHW021334080925
462592UK00007B/1

9 781784 275846